Innovative Strategies in Fisheries Nutrition

Sustainable Ingredients, Functional Additives, and Advanced Feeding Approaches for Aquaculture

NIPA® GENX ELECTRONIC RESOURCES & SOLUTIONS P. LTD.
New Delhi-110 034

About the Author

Dr. Ishtiyaq Ahmad earned his doctorate from the Fish Nutrition Research Laboratory, Department of Zoology, University of Kashmir, Srinagar, India. He is currently a Senior Researcher at the Division of Fish Genetics and Biotechnology, Faculty of Fisheries, Sher-e-Kashmir University of Agricultural Sciences & Technology Kashmir (SKUAST-K), Rangil, Ganderbal, India. His research focuses on developing pure lines for the genetic improvement of rainbow trout to enhance overall aquaculture production. Dr. Ahmad's expertise spans Fish Nutrition, Fish Physiology and Biochemistry, Nutrigenomics, Fish Parasitology, and Molecular Biology. His notable research includes establishing the optimal branched-chain amino acid requirements for rainbow trout. He has published over 40 research articles in esteemed international journals such as Scientific Reports, Aquaculture, Reviews in Aquaculture, Aquaculture Nutrition, Aquaculture Research, Aquaculture Reports, Journal of Applied Ichthyology, Frontiers in Marine Science, Food Chemistry: X, and Journal of Contaminant Hydrology. Additionally, Dr. Ahmad has authored several book chapters and five books with national and international publishers. He has presented his work at various state, national, and international conferences, seminars, and workshops, earning numerous accolades, including the Young Scientist Award from a national organization and several Best Paper Presentation awards. Dr. Ahmad serves on the editorial boards of prominent international journals, including PLOS ONE and Frontiers in Nutrition. He also worked as a Senior Research Fellow in a DBT-funded project on the nutrient requirements of rainbow trout fingerlings and is a recipient of the prestigious CSIR NET award.

Innovative Strategies in Fisheries Nutrition

Sustainable Ingredients, Functional Additives, and Advanced Feeding Approaches for Aquaculture

Ishtiyaq Ahmad
Senior Researcher
Division of Fish Genetics and Biotechnology
Faculty of Fisheries, Sher-e-Kashmir University of Agricultural Science and Technology, Kashmir, Handerbal
Jammu and Kashmir-190025, India

NIPA® GENX ELECTRONIC RESOURCES & SOLUTIONS P. LTD.
New Delhi-110 034

NIPA® GENX ELECTRONIC RESOURCES & SOLUTIONS P. LTD.

101,103, Vikas Surya Plaza, CU Block
L.S.C. Market, Pitam Pura, New Delhi-110 034
Ph : +91-11-43860225, Mob.: +91 9717133558, 9540816132
E-mail: newindiapublishingagency@gmail.com
Website: www.nipaersources.com

Print ISBN: 978-93-58874-59-4
ebook ISBN: 978-93-58878-91-2

Composed and Designed by NIPA®.

Government of Jammu & Kashmir
Higher Education Department

DEPARTMENT OF ZOOLOGY
SHMM, GOVT DEGREE COLLEGE ANANTNAG
Anantnag Khanbal-192101

Foreword

Aquaculture is fast becoming a cornerstone of global food security as the demand for high-quality, sustainable protein sources intensifies. Amid this critical shift, "*Innovative Strategies in Fisheries Nutrition: Sustainable Ingredients, Functional Additives and Advanced Feeding Approaches for Aquaculture*" offers a profound contribution to the science and practice of fish nutrition. This book brings together groundbreaking research and forward-thinking methodologies that promise to shape the future of the aquaculture industry. By addressing key challenges such as resource limitations, environmental sustainability, and the nutritional health of aquatic species, the book provides actionable insights for transforming fisheries nutrition.

The text explores a wide range of innovative approaches, from the use of functional feed additives and novel protein sources to the integration of nutrigenomics and precision feeding strategies. These techniques are not only enhancing growth performance and immunity in farmed fish but also reducing the ecological footprint of aquaculture. In addition, the focus on alternative protein sources and the sustainable use of raw materials ensures that aquaculture practices evolve in harmony with global environmental goals.

The book underscores the importance of holistic approaches to fish nutrition, advocating for strategies that prioritize both the economic viability and environmental sustainability of aquaculture systems. It highlights the significance of understanding species-specific nutritional needs and developing feeds that promote optimal health, growth, and resistance to disease. The introduction of bioactive compounds, probiotics, and prebiotics into fish diets is discussed as a promising avenue for boosting immunity and reducing reliance on synthetic chemicals.

This volume is a vital resource for researchers, nutritionists, aquaculture producers, and policymakers, offering them the tools to drive innovation and sustainability in fish nutrition. As the global aquaculture sector continues to expand, "*Innovative Strategies in Fisheries Nutrition: Sustainable Ingredients, Functional Additives and Advanced Feeding Approaches for Aquaculture*" provides essential guidance for navigating the complexities of meeting growing food demands while safeguarding the planet's natural resources. It is a bold and timely contribution to the field, setting the stage for a new era in sustainable aquaculture.

Sincerely,

Dr. Amir Maqbool

Assistant Professor
Department of Zoology

Preface

The rapid expansion of aquaculture as a solution to meet the rising global demand for seafood has underscored the critical need for innovative strategies in fish nutrition. Sustainable development in aquaculture depends not only on increasing production but also on ensuring environmental stewardship, enhancing fish welfare, and delivering high-quality, safe products to consumers. Innovative Strategies in Fisheries Nutrition: Sustainable Ingredients, Functional Additives, and Advanced Feeding Approaches for Aquaculture addresses these priorities by exploring sustainable feed solutions, cutting-edge ingredients, and functional additives that drive growth and resilience in aquaculture species.

This book presents a comprehensive overview of the latest advances in fish nutrition, focusing on the formulation of diets that optimize growth and health while reducing the ecological footprint of aquaculture operations. By examining sustainable protein and lipid sources, novel functional ingredients, and adaptive feeding methodologies, the book seeks to equip researchers, students, and professionals in aquaculture with actionable knowledge to enhance productivity and sustainability in the sector. Topics such as insect-based proteins, algae as a superfood, and functional feed additives highlight the exciting potential of alternative ingredients that can replace conventional fishmeal and fish oil, which are often unsustainable and costly.

As the aquaculture industry transitions from traditional feeding practices to approaches that incorporate bioactive compounds and precision nutrition, this book emphasizes the benefits of functional feed additives, including immunostimulants, prebiotics, and probiotics, which help improve fish health, reduce disease outbreaks, and lessen reliance on antibiotics. Advanced feeding strategies, such as phase-specific feeding, bioavailability optimization, and the use of nutrigenomics, are also discussed, providing a foundation for future research and practice that could transform aquaculture's nutritional standards.

Innovative Strategies in Fisheries Nutrition is intended as a resource for aquaculture scientists, nutritionists, policymakers, and students, offering insights into sustainable, efficient, and economically viable solutions for the nutrition of aquatic animals. I hope this book will contribute to the body of knowledge supporting sustainable aquaculture development and inspire further innovation and research in fisheries nutrition to meet the demands of a growing global population.

Author

Acknowledgement

The journey of writing ***"Innovative Strategies in Fisheries Nutrition: Sustainable Ingredients, Functional Additives and Advanced Feeding Approaches for Aquaculture"*** has been both challenging and rewarding, and it would not have been possible without the support of many individuals and organizations.

First and foremost, I would like to express my deepest gratitude to my family and friends. Your encouragement and belief in my vision have been my guiding light throughout this endeavor. Thank you for your patience during the long hours of research and writing, and for providing a nurturing environment that allowed my ideas to flourish. I extend my sincere appreciation to NIPA Publishers for their unwavering support and commitment to excellence in publishing. From the initial stages of manuscript preparation to the final publication, their professionalism and guidance have been instrumental in bringing this book to life. I am especially grateful for their collaborative spirit and responsiveness, which made the publishing process seamless and enjoyable.

I would also like to acknowledge the contributions of fellow researchers and practitioners in the field of fisheries nutrition. Their pioneering work has provided a rich foundation upon which this book is built. I am inspired by their dedication to advancing knowledge and practices in this vital area, and I hope to contribute to the ongoing dialogue through the insights shared in this book. Additionally, I want to recognize the mentors and colleagues who have shaped my understanding of fisheries nutrition and have offered invaluable advice throughout my academic and professional journey. Your insights and perspectives have been a source of inspiration and have enriched my work in countless ways.

Finally, I hope that Innovative Strategies in Fisheries Nutrition: Sustainable Ingredients, Functional Additives and Advanced Feeding Approaches for Aquaculture serves as a valuable resource for researchers, practitioners, and students alike. It is my aspiration that this book not only provides insights into innovative approaches in fisheries nutrition but also fosters a spirit of inquiry and collaboration among those dedicated to enhancing the sustainability and productivity of aquatic systems.

Thank you to everyone who has played a role in this journey. Your support has made this book possible.

Contents

1

Introduction to Innovative Approaches in Fisheries Nutrition

Introduction

As global populations increase, so does the demand for high-quality seafood. Aquaculture has emerged as a vital solution, contributing significantly to global fish production and food security. However, traditional practices in fisheries nutrition, primarily reliant on fishmeal and fish oil, face substantial challenges. Overfishing, rising feed costs, and environmental concerns necessitate a shift towards more sustainable and innovative approaches in aquaculture nutrition. This chapter delves into the innovative methodologies and technologies shaping the future of fisheries nutrition, emphasizing sustainability, health, efficiency, and the optimization of nutritional strategies.

1. The Need for Innovation in Fisheries Nutrition

1.1 Global Demand for Seafood

The world's appetite for seafood is growing, with aquaculture projected to meet a significant portion of this demand. According to the Food and Agriculture Organization (FAO), aquaculture accounts for over half of the fish consumed globally, making it essential to optimize production methods, including nutrition.

1.2 Traditional Feed Challenges

Historically, aquaculture feeds have relied heavily on fishmeal and fish oil derived from wild fisheries. These ingredients are high in protein and essential fatty acids but are not sustainable. The reliance on finite marine resources has led to overexploitation, with some fish stocks being classified as overfished or depleted. Furthermore, the cost of fishmeal and fish oil has been volatile, posing economic challenges for fish farmers.

1.3 Environmental Concerns

Aquaculture's environmental footprint is a growing concern. Nutrient runoff from fish farms can lead to water pollution and eutrophication in nearby aquatic systems. Hence, there is an urgent need to develop innovative feeding strategies that minimize waste and improve nutrient utilization while ensuring the health of aquatic ecosystems.

2. Sustainable Feed Ingredients

2.1 Plant-Based Proteins

Plant proteins, such as soybean meal, canola meal, and pea protein, have emerged as viable alternatives to fishmeal. These ingredients are abundant, renewable, and can be produced with a lower environmental impact. However, the challenge lies in their nutritional profile, which often lacks certain essential amino acids.

To address this, several strategies have been developed:

- **Formulation Optimization**: Combining various plant proteins can enhance the overall amino acid profile, ensuring that the nutritional needs of fish are met.
- **Processing Techniques**: Methods like extrusion, fermentation, and hydrolysis can improve the digestibility and nutritional value of plant-based feeds. For instance, fermentation can reduce anti-nutritional factors and enhance the bioavailability of nutrients.

2.2 Insect Meal

Insect protein is gaining recognition as a sustainable alternative to traditional fishmeal. Insects such as black soldier flies and mealworms are rich in protein and essential fatty acids, making them suitable for various fish species. Their production is also environmentally friendly, as they can be raised on organic waste, thereby reducing food waste and greenhouse gas emissions.

2.3 Algal Ingredients

Microalgae are increasingly incorporated into aquafeeds due to their high omega-3 fatty acid content (EPA and DHA) and their role in promoting fish health. Algal oils can serve as a direct substitute for fish oil, reducing reliance on marine resources. Furthermore, algae can provide additional nutrients, such as antioxidants, vitamins, and minerals, which contribute to improved fish health and growth.

2.4 Single-Cell Proteins (SCP)

SCP derived from bacteria, yeast, and fungi are emerging as innovative feed ingredients. These microorganisms can be cultivated on various substrates, including agricultural by-products, making their production sustainable and efficient. SCPs offer high protein content and can be tailored to meet specific nutritional requirements of different fish species.

3. Functional Feed Additives

Functional feed additives enhance fish health and performance beyond basic nutrition. These additives can improve growth rates, boost immunity, and enhance feed efficiency, contributing to the sustainability of aquaculture systems.

3.1 Probiotics and Prebiotics

Probiotics are live beneficial microorganisms that can confer health benefits when administered in adequate amounts. They improve gut health, nutrient absorption, and immune responses in fish. Prebiotics, on the other hand, are non-digestible food components that stimulate the growth of beneficial gut bacteria. Together, these additives can enhance the overall health of fish and reduce the need for antibiotics in aquaculture.

3.2 Immunostimulants

Immunostimulants are compounds that enhance the immune response in fish, making them more resilient to diseases. Natural substances such as beta-glucans, garlic extracts, and certain vitamins can be incorporated into feeds to stimulate immune function. For instance, beta-glucans derived from yeast have been shown to enhance the immune response and disease resistance in various fish species.

3.3 Phytogenics

Phytogenic compounds, derived from plants, possess bioactive properties that can improve fish health. These compounds include essential oils, herbs, and spices, which have antioxidant, antimicrobial, and anti-inflammatory properties. By incorporating phytogenics into aquafeeds, producers can enhance growth performance, improve feed efficiency, and support the overall health of farmed fish.

3.4 Omega-3 Fatty Acids

Supplementing aquafeeds with omega-3 fatty acids from alternative sources, such as algae and by-products of fish processing, is crucial for maintaining the health of carnivorous fish species. These fatty acids play vital roles in cellular function, growth, and overall health. Ensuring adequate omega-3 intake can lead to improved growth rates and better overall health in fish.

4. Precision Nutrition and Feeding Technologies

Precision nutrition involves tailoring diets to the specific nutritional needs of fish, considering factors such as species, life stage, and environmental conditions. This approach optimizes feed efficiency, minimizes waste, and enhances overall sustainability.

4.1 Nutrigenomics

Nutrigenomics is the study of the interaction between nutrition and gene expression. By understanding how specific nutrients influence the genetic makeup of fish, researchers can develop targeted feeding strategies that optimize growth and health. For example, identifying genetic markers associated with nutrient utilization can inform feed formulations that cater to the specific needs of different fish species.

4.2 Automated Feeding Systems

Automated feeding systems equipped with sensors and artificial intelligence (AI) technology are transforming the way fish are fed. These systems allow for precise control of feeding schedules and quantities based on real-time data on fish behavior and environmental conditions. By reducing feed waste and ensuring that fish receive the right amount of nutrition, automated systems contribute to improved feed conversion ratios and overall farm efficiency.

4.3 Smart Aquafeeds

Smart aquafeeds are designed to deliver nutrients in a targeted manner, ensuring that fish receive the right amount of nutrition at the right time. These feeds may use microencapsulation or controlled-release technologies to protect sensitive nutrients from degradation during storage and processing. By enhancing nutrient bioavailability and reducing waste, smart aquafeeds can significantly improve the efficiency of aquaculture systems.

5. Advanced Feed Formulation and Processing Techniques

Innovative feed formulation and processing techniques are critical to enhancing the nutritional quality, digestibility, and palatability of aquafeeds. These advancements contribute to the development of highly efficient feeds that cater to the specific needs of different fish species.

5.1 Fermentation

Fermentation is a process that can improve the nutritional value of plant-based ingredients by breaking down complex nutrients and reducing anti-nutritional factors. Fermented feeds have been shown to enhance growth performance and nutrient digestibility in fish. For example, fermented soybean meal has been successfully used as a protein source in aquafeeds, improving its nutritional profile.

5.2 Microencapsulation

Microencapsulation is a technique that protects sensitive nutrients from degradation during feed processing and storage. This technology allows for the controlled release of nutrients, ensuring that fish receive the necessary vitamins, minerals, and essential fatty acids when they are most needed. Microencapsulated feeds have been shown to improve nutrient utilization and overall fish performance.

5.3 Enzyme Supplementation

The inclusion of exogenous enzymes in aquafeeds enhances the digestibility of plant-based proteins and carbohydrates. Enzymes such as phytase, protease, and amylase can break down complex nutrients into more readily available forms, improving feed efficiency and growth performance. For instance, the addition of phytase can help reduce phosphorus excretion, minimizing the environmental impact of aquaculture.

6. Role of Innovation in Addressing Future Challenges

The innovative approaches in fisheries nutrition outlined in this chapter are instrumental in addressing the future challenges facing global aquaculture. These challenges include resource scarcity, environmental degradation, and the need for increased efficiency in production systems.

6.1 Enhancing Resource Efficiency

By developing sustainable feed ingredients and optimizing feed formulations, aquaculture can reduce its reliance on finite marine resources. Innovations such as insect meal, plant-based proteins, and algal ingredients offer viable alternatives that not only support fish health but also promote environmental sustainability.

6.2 Reducing Environmental Impact

Innovative feeding strategies can significantly reduce the environmental footprint of aquaculture. Precision nutrition and automated feeding systems minimize feed waste and nutrient runoff, while functional additives improve fish health and resilience, reducing the need for antibiotics and other treatments.

6.3 Supporting Food Security

As the global demand for seafood continues to rise, the aquaculture industry must be equipped to meet this demand sustainably. Innovations in fisheries nutrition will enhance the productivity and efficiency of aquaculture systems, ensuring a stable supply of high-quality seafood that supports global food security.

Table: Overview of Innovative Approaches in Fisheries Nutrition

Topic	Descriptions	Examples/Innovations
Global Demand for Seafood	The increasing need for seafood to meet the nutritional requirements of a growing global population.	Aquaculture producing over half of global fish consumption.
Challenges of Traditional Feeds	Reliance on fishmeal and fish oil leading to sustainability concerns, high costs, and environmental impact.	Overfishing and depletion of marine resources.
Sustainable Feed Ingredients	Exploring alternative ingredients that are renewable and environmentally friendly.	- Plant-Based Proteins: Soybean meal, pea protein. - Insect Meal: Black soldier flies, mealworms. - Algal Ingredients: Omega-3 rich microalgae. - Single-Cell Proteins: Bacteria, yeast, fungi.
Functional Feed Additives	Enhancing health and performance of fish through bioactive components.	- Probiotics & Prebiotics: Promote gut health. - Immunostimulants: Beta-glucans, garlic extracts. - Phytogenics: Essential oils from plants. - Omega-3 Fatty Acids: Supplementing with algal oils.
Precision Nutrition	Tailoring diets to the specific nutritional needs of fish based on species and conditions.	- Nutrigenomics: Genetic influence on nutrient absorption. - Automated Feeding Systems: Real-time data control. - Smart Aquafeeds: Controlled release of nutrients.
Advanced Feed Formulation	Techniques to enhance the digestibility and nutritional quality of aquafeeds.	- Fermentation: Improved nutrient profile through microbial activity. - Microencapsulation: Protects sensitive nutrients. - Enzyme Supplementation: Enhances nutrient digestibility.
Future Challenges in Aquaculture	Addressing issues like resource scarcity, environmental degradation, and food security.	Innovations supporting sustainable aquaculture practices.

Conclusion

Innovative approaches in fisheries nutrition are reshaping the aquaculture industry, offering sustainable, efficient, and health-promoting solutions to meet the growing global demand for seafood. From the development of alternative feed ingredients to precision feeding technologies and functional additives, these advancements are paving the way for a more resilient and responsible aquaculture sector. The continued adoption of these innovations will be essential for ensuring the future sustainability of aquaculture and its ability to contribute to global food security. By embracing these innovative strategies, the aquaculture industry can navigate the challenges of the 21st century, ultimately fostering a more sustainable and equitable food system for all.

References

Barlow, J. W. (2017). Advancements in Fish Nutrition: The Role of Enzyme Supplementation in Aquaculture. Aquaculture Nutrition, 23(5), 1010-1017.

Barroso, F. G., et al. (2014). Insect Meal as a Protein Source in Fish Feed: An Overview. Aquaculture Research, 45(4), 609-617.

Coyle, S. D., et al. (2015). Automated Feeding Technologies in Aquaculture. Aquaculture Engineering, 64, 1-10.

Dawes, C. J. (2018). Marine Algae and Aquaculture: The Role of Seaweed in Aquafeeds. CABI.

Enright, A., et al. (2020). Nutrigenomics in Aquaculture: Approaches and Applications. Aquaculture, 528, 735455.

Food and Agriculture Organization (FAO). (2020). The State of World Fisheries and Aquaculture 2020: Sustainability in Action. FAO.

Gatlin, D. M., Barrows, F. T., Brown, P., Dabrowski, K., & De Marco, A. (2007). Expanding the Utilization of Sustainable Plant Products in Aquafeeds: A Review. Aquaculture Research, 38(6), 551-579.

Gjedrem, T., & Baranski, M. (2010). Selective Breeding in Aquaculture: An Introduction. CABI Publishing.

Glencross, B. D. (2006). The Nutrition of Rainbow Trout: A Review. Aquaculture Nutrition, 12(6), 449-460.

Krogdahl, Å., et al. (2010). The Role of Dietary Fiber in Fish Nutrition: Perspectives on the Use of Plant Ingredients in Aquaculture. Aquaculture Research, 41(2), 161-182.

Lee, S. M., & Coyle, S. D. (2011). Insect Protein: A Novel Ingredient in Aquaculture Feed. Aquaculture Magazine, 37(1), 14-18.

Naylor, R. L., & Burke, M. (2005). Aquaculture and Ocean Resources: Raising Tigers of the Sea. Annual Review of Environment and Resources, 30, 185-218.

Romero, J., et al. (2015). Functional Feed Additives: Probiotics, Prebiotics, and Their Applications in Aquaculture. Fish & Shellfish Immunology, 45(1), 1-5.

Tacon, A. G. J., & Metian, M. (2013). Feed Matters: Satisfying the Feed Demand of Aquaculture. Fish Physiology and Biochemistry, 39(2), 439-455.

Yi, Y., et al. (2013). The Role of Probiotics and Prebiotics in Fish Nutrition. Journal of Applied Ichthyology, 29(6), 1035-1043.

2

Nutritional Physiology of Fish A Foundation for Innovation

Introduction

Fish nutrition is a critical determinant of growth, reproduction, and overall health in aquaculture systems. A thorough understanding of the nutritional physiology of fish lays the groundwork for designing innovative feeding strategies that improve fish production, health, and sustainability. Unlike terrestrial animals, fish have unique physiological traits that influence their dietary needs and how they process nutrients, requiring specialized nutritional approaches. This chapter provides a detailed exploration of these physiological features and their implications for the development of cutting-edge nutritional innovations in aquaculture.

1. Basics of Fish Anatomy and Physiology Related to Nutrition

Fish anatomy and physiology particularly in the context of nutrition present unique characteristics that differentiate them from terrestrial animals. One of the most significant features of fish is their poikilothermic nature, meaning their body temperature fluctuates with the surrounding water. This directly impacts their metabolism, as colder waters slow down metabolic processes, reducing the need for energy, while warmer waters increase metabolic rates, driving higher energy demand. This temperature dependency is critical in understanding the feeding behavior and nutritional needs of different fish species. Unlike mammals or birds, which expend energy to maintain a constant internal body temperature, fish can allocate a greater proportion of their consumed nutrients toward growth, reproduction, and immune function.

Another key aspect of fish biology related to nutrition is their specialized digestive systems, which vary significantly between species based on their diets. Carnivorous fish, like salmon and trout, typically have short, simple digestive tracts designed to rapidly digest protein-rich meals. In contrast, herbivorous fish, such as tilapia, have longer, more complex intestines that allow for the fermentation and breakdown of plant matter, which is more difficult to digest. Omnivorous species like carp fall somewhere in between, having digestive systems capable of processing both plant and animal materials. Fish digest their food through a coordinated process involving several key organs. The stomach, present in most fish species, secretes enzymes and acids that initiate the breakdown of proteins. The pancreas plays a critical role in secreting digestive enzymes that further aid in breaking down fats, proteins, and carbohydrates. The liver is essential for metabolism and energy storage, processing nutrients and distributing them throughout the body. Additionally, the intestines are where the majority of nutrient absorption takes place. In some species, the presence of pyloric caeca (finger-like extensions near the stomach) enhances nutrient absorption by increasing the surface area.

Water plays a unique role in fish digestion and nutrient absorption. Because fish live in water, they are continuously surrounded by a medium that influences osmoregulation, or the balance of salts and water within their bodies. Minerals like sodium, potassium, and chloride are absorbed from the water, aiding in osmoregulation and impacting how fish process their food. For instance, fish living in freshwater environments must work to retain salts and expel excess water, whereas marine fish face the opposite challenge, needing to expel excess salts while retaining water. This osmotic balance affects the way nutrients are absorbed and processed, adding another layer of complexity to fish nutrition.

Furthermore, fish nutrition is heavily influenced by environmental factors, such as oxygen levels, water quality, and availability of food. Poor water quality, especially in intensive aquaculture systems, can stress fish and reduce their ability to digest and absorb nutrients efficiently. Low oxygen levels can decrease the energy available for metabolic processes, leading to slower growth and reduced feed conversion efficiency. Therefore, maintaining optimal water quality and ensuring that the diet matches the environmental conditions is critical for enhancing fish health and performance.

In terms of energy utilization, fish derive most of their energy from proteins and lipids. Protein is particularly important in carnivorous fish, which require a high intake to maintain tissue development and repair. Lipids provide a dense source of energy, and many fish species also require essential fatty acids like omega-3 and omega-6, which are crucial for cell membrane integrity, immune function, and growth. Carbohydrates while less important in the diet of carnivorous species, can serve as an important energy source for herbivorous and omnivorous fish, helping to spare protein for growth rather than using it for energy.

The nutritional physiology of fish also includes the role of vitamins and minerals. Vitamins such as A, D, E, and K, along with water-soluble vitamins like B-complex, are crucial for proper metabolic functions, including growth, reproduction, and immunity. For example, vitamin C is essential for collagen synthesis, wound healing, and immune response, and a deficiency can lead to impaired growth and increased susceptibility to disease. Minerals such as calcium and phosphorus are critical for skeletal development and metabolic processes. Fish in natural habitats may acquire these nutrients from the water as well as their diet, while in aquaculture; their feed must be carefully formulated to meet these essential requirements.

In recent years, the field of fish nutrition has been revolutionized by advancements in functional feeds, designed to go beyond basic nutrition to promote health, disease resistance, and enhanced growth. These feeds often incorporate probiotics, prebiotics, and other bioactive compounds that improve gut health and immune response. For instance, the addition of probiotics like *Lactobacillus* can help modulate the gut microbiota, improving nutrient absorption and reducing pathogenic bacteria. Phytogenic additives, such as essential oils from herbs like ginger and garlic, have been shown to enhance digestion and boost the immune system. These functional additives are particularly beneficial in intensive aquaculture systems, where fish are more prone to stress and disease due to high stocking densities and environmental pressures.

Understanding fish nutritional physiology is essential for optimizing feeding strategies in aquaculture, ensuring that fish receive the right balance of nutrients for their specific species, life stage, and environmental conditions. By tailoring diets to meet these needs, aquaculture operations can improve growth rates, feed conversion ratios, and overall fish health. Additionally, with the growing focus on sustainability in aquaculture, research into alternative feed ingredients, such as plant-based proteins, insect meals, and microalgae, is becoming increasingly important. These sustainable alternatives can help reduce reliance on traditional fishmeal and fish oil, which are derived from wild fish stocks and contribute to overfishing concerns.

1.1 Digestive Anatomy

The digestive system in fish can vary widely depending on species, particularly between carnivorous, herbivorous, and omnivorous species. The primary organs involved in fish digestion include the mouth, esophagus, stomach, intestines, liver, pancreas, and gallbladder.

- **Mouth and Esophagus:** These serve as the initial points of food intake. The morphology of the mouth, including the presence of teeth and the structure of the esophagus, is often adapted to the feeding habits of the species. For instance, carnivorous fish like bass have well-developed teeth to capture and consume prey, while herbivorous fish like tilapia possess fine teeth suitable for grazing on plant matter.

- **Stomach:** Carnivorous species, such as salmon, possess a stomach that secretes digestive enzymes and hydrochloric acid to break down proteins and fats. Herbivorous species may have a less developed or even absent stomach, instead relying more on the intestine for digestion.
- **Intestines:** The length of the intestine is directly related to the diet of the fish. Herbivores typically have longer intestines to accommodate the fermentation and breakdown of plant materials, whereas carnivores have shorter intestines for rapid digestion of protein-rich prey. For example, the gastrointestinal tract of a tilapia can be 2-3 times longer than that of a carnivorous trout.
- **Liver and Pancreas:** The liver and pancreas are critical for producing digestive enzymes and bile, aiding in the breakdown of fats, proteins, and carbohydrates. The liver also plays a key role in nutrient metabolism and storage.

Understanding these anatomical differences is essential for designing diets tailored to the specific digestive capacities of different fish species.

2. Macronutrient Requirements in Fish

Macronutrients, including proteins, lipids, and carbohydrates, are the primary nutritional components required by fish for energy, growth, and maintenance. Their intake and metabolism differ from those of terrestrial animals due to fish physiology and aquatic living conditions.

2.1 Protein

Proteins are the most critical macronutrient for fish, primarily due to their essential role in growth, tissue repair, enzyme production, and overall metabolic functions. They serve as the building blocks for the development of muscle tissue, organs, and cellular structures. In addition to structural functions, proteins are integral in biochemical processes, functioning as enzymes that catalyze metabolic reactions, hormones that regulate physiological functions, and antibodies that support the immune system. Fish are particularly efficient in utilizing proteins as an energy source, often at the expense of carbohydrates or lipids, which makes protein the cornerstone of their nutritional requirements.

- **Amino Acid Requirements:** Amino acids are the fundamental units of proteins, and fish require a specific balance of essential amino acids that must be obtained from their diet because they cannot synthesize them internally. Essential amino acids for fish include lysine, methionine, threonine, arginine, histidine, leucine, isoleucine, and valine. The balance and availability of these amino acids are crucial for optimal growth, reproduction, and overall health. For instance, lysine and methionine are often the first limiting amino acids in fish diets, meaning that inadequate levels of these amino acids can restrict growth, impair feed utilization, and lead to suboptimal performance. In carnivorous species, which have higher protein requirements due to their natural diets, maintaining an appropriate amino acid profile in their feed is even more critical. This is particularly important during the early life stages of fish, where rapid growth and development demand a higher intake of essential amino acids.

- **Protein Quality:** The quality of protein sources in aquaculture feeds is paramount for ensuring that fish receive the essential amino acids needed for growth and health. Traditionally, fishmeal has been the primary source of protein in aquaculture diets, prized for its high digestibility and rich amino acid profile, closely matching the nutritional needs of many fish species. However, environmental and economic concerns regarding the overfishing of wild stocks have prompted a search for sustainable alternatives. As a result, alternative protein sources such as plant proteins—specifically soybean meal, corn gluten meal, and pea protein—have gained significant attention in recent years. While these plant-based proteins can provide substantial quantities of protein, they often come with limitations in terms of amino acid profiles and digestibility. For example, soybean meal is high in protein but low in methionine, which can be a challenge for formulating diets that meet the complete nutritional needs of fish. In addition to plant proteins, recent innovations have introduced insect meal as a sustainable, protein-rich alternative. Insect larvae, such as black soldier fly larvae and mealworms, have demonstrated high protein content and a favorable amino acid profile that can complement fish diets. Moreover, insect meal is produced with lower environmental footprints compared to traditional fishmeal, offering an eco-friendly option for aquaculture.
- **Digestibility and Metabolism:** Fish possess remarkable digestive systems that allow them to metabolize protein efficiently, with protein digestibility rates often exceeding 90%. However, the digestibility of protein can vary significantly depending on factors such as the species of fish, the water temperature, and the source of protein. For example, marine species, such as tuna and salmon, typically have higher protein requirements due to their faster growth rates and metabolic pathways,

which are more geared toward protein utilization. Conversely, freshwater species like carp have different metabolic adaptations that may allow for a greater proportion of their diet to consist of carbohydrates and lipids, affecting their overall protein requirements. The temperature of the water also plays a vital role in protein metabolism. Higher temperatures tend to enhance metabolic rates, leading to increased protein requirements for growth and maintenance, whereas lower temperatures can slow down metabolism, resulting in reduced protein needs. Moreover, the efficiency of protein utilization can be influenced by the form in which proteins are presented in feeds. The physical form (pellet, crumble, or powder) and the processing methods (such as cooking or fermentation) can enhance protein digestibility and nutrient availability. For instance, enzymatic treatments that break down complex proteins into smaller peptides may improve the absorption rates in fish intestines.

2.2 Lipids

Lipids are a crucial macronutrient in fish nutrition, playing multiple essential roles that significantly impact fish health, growth, and overall well-being. They provide a concentrated energy source, facilitate the absorption of fat-soluble vitamins, and serve as precursors for essential fatty acids and hormones. As an energy-dense nutrient, lipids contain approximately twice the energy per gram compared to carbohydrates and proteins, making them vital for meeting the energy demands of fish, particularly in aquaculture.

- **Energy and Fatty Acid Metabolism:** Lipids serve as the primary energy source for many fish species, especially carnivorous fish, which often have limited carbohydrate metabolism. Unlike terrestrial animals, fish have adapted to utilize lipids efficiently as their main energy substrate, particularly during periods of high energy demand, such as rapid growth phases, reproduction, and recovery from stress. This unique metabolic capability allows for a protein-sparing effect, meaning that when sufficient lipid energy is available, fish can redirect protein usage toward growth and tissue repair rather than using it solely for energy production. In aquaculture diets, lipids typically constitute 10-20% of the feed, with the specific percentage varying depending on the species and life stage of the fish. For instance, carnivorous species like salmon may require higher lipid levels in their diets compared to omnivorous or herbivorous species. The types of lipids included in the diet can also influence fish health; for example, lipid sources rich in long-chain polyunsaturated fatty acids (LC-PUFAs) are particularly beneficial for growth and overall health.

- **Essential Fatty Acids:** Among the lipids, essential fatty acids are of particular importance to fish nutrition. Long-chain polyunsaturated fatty acids (LC-PUFAs), especially eicosapentaenoic acid (EPA) and docosahexaenoic acid (DHA), are critical for maintaining fish health. These omega-3 fatty acids are involved in various physiological processes, including immune response, cell membrane integrity, and anti-inflammatory actions. EPA and DHA play essential roles in the development of the brain and nervous system in fish, which is particularly important during the early life stages when rapid growth and organ development occur. Marine fish, such as salmon and mackerel, have higher dietary requirements for omega-3 fatty acids due to their natural diets, which are rich in these fatty acids from prey items like small fish and krill. Conversely, freshwater species may have a greater reliance on plant-based sources of omega-6 fatty acids, which can be derived from sources like corn oil and soybean oil. The balance of omega-3 to omega-6 fatty acids in the diet is crucial for ensuring optimal health, growth, and reproduction in fish.
- **Innovations in Lipid Sources:** Traditionally, fish oil has been the primary source of essential fatty acids in aquaculture feeds, largely due to its high levels of EPA and DHA. However, concerns over overfishing and the sustainability of fish oil production have driven research into alternative lipid sources. Various innovative lipid sources are now being explored to meet the nutritional requirements of fish while promoting environmental sustainability.
    1. **Algae-Based Oils**: Algal oil is emerging as a sustainable alternative to fish oil, rich in DHA and other essential fatty acids. Algae can be cultivated in controlled environments, reducing the pressure on marine ecosystems. Algal oil is increasingly being used in the production of feeds for high-value species like salmon and trout, providing a viable means of meeting their omega-3 fatty acid requirements without depleting fish stocks.
    2. **Vegetable Oils**: Various vegetable oils, such as soybean oil, canola oil, and palm oil, are being incorporated into aquaculture feeds to provide energy and essential fatty acids. While vegetable oils may have lower levels of EPA and DHA compared to fish oil, they can still contribute to the overall lipid profile of the diet and offer a more sustainable option.
    3. **Microbial Lipids**: The use of microbial lipids derived from single-cell organisms, such as yeast and fungi, is gaining traction as another

innovative source of lipids. These microbial sources can be cultivated on agricultural byproducts, allowing for the production of omega-3 and omega-6 fatty acids while reducing competition with food crops.

4. **Enrichment Techniques**: Innovations in lipid enrichment techniques are also being explored, such as supplementing fish feeds with omega-3 fatty acids through direct addition or by using microencapsulation techniques to enhance the stability and availability of these fatty acids during digestion.

2.3 Carbohydrates

Carbohydrates are one of the three primary macronutrients essential for the growth and health of fish, yet they are the least utilized in fish diets compared to proteins and lipids. This limited utilization is especially evident in carnivorous species that possess a reduced capacity to digest and metabolize complex carbohydrates. The unique anatomical and physiological characteristics of fish, particularly their digestive systems, play a crucial role in how different species interact with carbohydrates.

- **Carbohydrate Metabolism:** The metabolism of carbohydrates in fish varies significantly based on their feeding habits. Carnivorous fish, such as salmon and trout, typically have lower levels of amylase and other carbohydrate-digesting enzymes in their digestive tracts. This enzymatic limitation restricts their ability to efficiently break down and utilize carbohydrates, making them less dependent on this macronutrient for energy. In contrast, herbivorous and omnivorous fish species, such as carp and tilapia, exhibit a greater ability to utilize carbohydrates. These species possess longer intestines, allowing for extended digestion and absorption of nutrients. Additionally, the gut microbiota of these fish plays a significant role in carbohydrate metabolism. Specialized bacteria in the intestines can ferment plant-based carbohydrates, breaking them down into simpler sugars that can be absorbed and utilized for energy. This differential carbohydrate metabolism underscores the importance of formulating species-specific diets in aquaculture. Understanding the dietary requirements of various fish species is essential for optimizing growth performance and feed efficiency.
- **Energy Source:** In many aquaculture systems, carbohydrates are included in fish feeds primarily as a cost-effective source of energy. Their inclusion helps spare protein from being used solely for energy production, allowing more amino acids to be directed toward growth and tissue development. This protein-sparing effect is particularly beneficial in high-protein diets, where ensuring adequate protein levels is critical

for optimal growth. The proportion of carbohydrates in aquaculture feeds typically ranges from 20-40%, depending on the species and their life stage. For instance, herbivorous species may require higher carbohydrate levels in their diets, while carnivorous species may benefit from lower carbohydrate inclusion and higher protein and lipid content. Additionally, the type and source of carbohydrates can influence growth performance and feed efficiency, highlighting the need for careful diet formulation.

- **Innovations in Carbohydrate Use:** Recent advancements in feed technology have opened new avenues for improving carbohydrate utilization in fish diets. Some of these innovations include:

  1. **Enzymatic Supplements**: The incorporation of enzymes such as amylase and cellulose in aquaculture feeds can enhance the digestibility of plant-based carbohydrates. These enzymes facilitate the breakdown of complex carbohydrates into simpler sugars, making them more accessible for absorption by the fish. This approach can significantly improve feed conversion ratios, leading to better growth performance.
  2. **Fermentation Processes**: Utilizing fermentation processes to pre-treat feed ingredients can enhance their nutritional value. Fermentation can break down anti-nutritional factors present in some plant materials, improving their digestibility. Additionally, fermentation can produce beneficial metabolites, such as organic acids, which can positively influence gut health and nutrient absorption in fish.
  3. **Gelatinized Starches**: The development of gelatinized starches represents a significant innovation in carbohydrate utilization. Gelatinization involves the modification of starches through heat and moisture, making them more digestible for fish. This process improves the availability of carbohydrates, especially for carnivorous species that may struggle with complex starches. Gelatinized starches can lead to more efficient energy metabolism and reduced feed costs in aquaculture.
  4. **Alternative Carbohydrate Sources**: The exploration of alternative carbohydrate sources, such as seaweeds, insects, and other unconventional feed ingredients, offers exciting possibilities for enhancing the nutritional profile of aquaculture feeds. These alternative sources may contain unique carbohydrate structures that can be beneficial for specific fish species, leading to improved growth rates and overall health.

3. Micronutrients in Fish Nutrition

Micronutrients, encompassing vitamins and minerals, are vital components of fish nutrition that play essential roles in various metabolic processes, immune function, growth, and overall health. Although required in smaller quantities than macronutrients, their significance in maintaining fish physiology and ensuring optimal health cannot be overstated. A well-balanced diet that includes the appropriate levels of micronutrients is essential for promoting health, growth, and productivity in aquaculture systems.

3.1 Vitamins

Vitamins are organic compounds that are necessary for numerous physiological functions in fish, including growth, reproduction, and resistance to diseases. They can be broadly categorized into two groups: fat-soluble and water-soluble vitamins.

Fat-Soluble Vitamins: Fat-soluble vitamins (A, D, E, and K) are absorbed along with dietary fats and stored in the fatty tissues and liver of fish. Each of these vitamins has unique and critical roles:

- **Vitamin A**: Vitamin A is crucial for vision, immune response, and growth. It is integral in the synthesis of visual pigments, which are essential for fish to see in low-light conditions. Additionally, vitamin A plays a significant role in cellular differentiation and proliferation, influencing growth rates and development. Deficiencies in vitamin A can lead to poor growth, increased susceptibility to infections, and impaired vision, making it essential for maintaining healthy populations in aquaculture.
- **Vitamin D**: This vitamin is vital for regulating calcium and phosphorus metabolism, both of which are critical for bone development and overall growth. Vitamin D aids in the absorption of these minerals from the digestive tract and ensures their availability for physiological processes. A deficiency in vitamin D can lead to skeletal deformities and growth retardation, emphasizing the need for adequate dietary levels.
- **Vitamin E**: Known for its potent antioxidant properties, vitamin E protects cell membranes from oxidative damage caused by reactive oxygen species. This vitamin is also crucial for immune function and reproductive health, supporting the integrity of reproductive tissues and enhancing overall health. Insufficient vitamin E can lead to reproductive failures and increased mortality in developing embryos.
- **Vitamin K**: Vitamin K plays an essential role in blood clotting and bone health. It is involved in synthesizing proteins required for coagulation

and maintaining bone structure. Deficiencies in vitamin K can lead to bleeding disorders and skeletal problems, underscoring its importance in fish health.

Water-Soluble Vitamins

Water-soluble vitamins, including the B vitamins (e.g., B12, B6, niacin, riboflavin) and vitamin C, are crucial for various metabolic functions and need to be regularly supplied in the diet since they are not stored in large amounts:

- **Vitamin C (Ascorbic Acid)**: Vitamin C is critical for collagen synthesis, immune response, and serving as an antioxidant. Fish, unlike many terrestrial animals, cannot synthesize vitamin C, making dietary supplementation necessary to prevent deficiencies. Insufficient levels of vitamin C can lead to poor growth, weakened immune responses, and increased susceptibility to diseases, particularly in juvenile fish.

3.2 Minerals

Minerals are inorganic elements that play key roles in fish osmoregulation, bone development, and enzymatic reactions. Essential minerals include both macro-minerals (e.g., calcium, phosphorus, magnesium) and trace minerals (e.g., zinc, iron, copper).

- **Calcium and Phosphorus:** These minerals are critical for bone development, growth, and metabolic functions. Calcium is acquired not only from the diet but also from the surrounding water, depending on the species and water chemistry. Phosphorus must be supplied through the diet and is vital for energy metabolism, serving as a component of adenosine triphosphate (ATP), the energy currency of the cell. Adequate levels of both minerals are necessary to ensure optimal growth rates and skeletal integrity.
- **Osmoregulation:** Maintaining proper electrolyte balance is crucial for fish to regulate the uptake and excretion of water. Minerals such as sodium, potassium, and chloride are essential for this process. Marine species, which live in hypertonic environments, must constantly excrete excess salt to maintain osmotic balance. An imbalance in these electrolytes can lead to stress, reduced growth rates, and compromised immune function.
- **Trace Elements:** Trace elements, including zinc, copper, manganese, and selenium, are required in small amounts but are essential for numerous enzymatic and antioxidant functions. For example, zinc plays a significant role in immune function and cellular metabolism, while selenium is a key component of antioxidant enzymes, such as glutathione

peroxidase, which protects fish from oxidative stress. Adequate intake of trace elements is crucial for maintaining overall health and enhancing growth performance in aquaculture.

4. Feeding Behavior and Energy Metabolism in Fish

Understanding the feeding behavior and energy metabolism of fish is crucial for optimizing their growth, health, and productivity in aquaculture. Fish exhibit diverse feeding strategies influenced by various factors, including species, water temperature, availability of food, and life stage. Recognizing these behaviors allows aquaculturists to tailor feeding regimes to better meet the nutritional needs of different fish species, ultimately enhancing their performance.

4.1 Continuous Feeders

Continuous feeders are species that engage in constant foraging behavior throughout the day. This group primarily includes herbivorous and omnivorous fish, such as tilapia and carp. These species have evolved to graze on algae, plant materials, and detritus, allowing them to derive nutrients from a wide variety of sources. In aquaculture settings, continuous feeders benefit from feeding strategies that mimic their natural behavior. Providing frequent, small meals throughout the day enhances nutrient absorption and minimizes waste. This feeding strategy aligns with the natural foraging behavior of these species, ensuring that they can efficiently utilize the available nutrients. Continuous feeding also supports optimal growth rates and overall health, as it reduces the likelihood of stress associated with prolonged fasting. Moreover, the design of aquaculture feeding systems for continuous feeders often incorporates automated feeders, allowing for precise control over feeding frequency and portion sizes. This technology not only optimizes feed utilization but also helps maintain water quality by reducing uneaten feed and waste, which can lead to detrimental effects on fish health and the aquatic environment.

4.2 Intermittent Feeders

Intermittent feeders, predominantly carnivorous species like salmon and trout, exhibit a different feeding strategy. These fish typically consume large meals at once and then undergo extended fasting periods. This feeding behavior has evolved as a survival mechanism, allowing these species to capitalize on sporadic food availability in their natural habitats. In aquaculture, intermittent feeders are commonly fed 2-3 times per day to simulate their natural feeding behavior. The size and frequency of meals are carefully managed to ensure that the fish can efficiently digest and absorb the nutrients from these larger feedings. Providing ample time between meals allows for complete digestion and reduces the risk of overfeeding, which can lead to increased waste and negatively impact water quality. The energy metabolism of intermittent feeders is adapted to utilize the energy derived from large meals effectively. The fish's digestive systems are specialized to handle significant protein intake, converting it into energy for growth and other physiological processes. Understanding the digestive capacity and energy requirements of these species is crucial for formulating diets that meet their nutritional needs and promote optimal growth rates.

4.3 Periodic Feeders

Some fish species, such as Arctic char and many migratory species, exhibit periodic feeding behavior, which is often influenced by environmental cues, including changes in daylight and seasonal availability of food. For instance, Arctic char may feed more aggressively during summer months when food is plentiful and available, while reducing their feeding activity in winter when resources are scarce. In aquaculture systems, recognizing the periodic feeding patterns of such species allows for adjustments to feeding schedules to align with their natural rhythms. By adapting feeding times and quantities based on environmental cues, aquaculturists can ensure that fish receive adequate nutrition during critical growth periods. This approach not only optimizes growth rates but also enhances the overall health and well-being of the fish. Additionally, periodic feeders may require specific dietary formulations during different life stages or seasons. For example, diets rich in certain nutrients may be provided during periods of rapid growth or before spawning, whereas maintenance diets with lower nutrient density may be adequate during less active seasons. Tailoring feeding strategies based on the periodic feeding behavior of fish enhances nutrient utilization and promotes sustainable aquaculture practices.

5. Innovative Approaches in Fish Nutrition Based on Physiology

In recent years, advancements in fish physiology and nutrition have led to innovative feeding strategies designed to meet the specific physiological needs of fish in aquaculture. These approaches not only aim to enhance growth and health but also improve sustainability and feed efficiency, aligning with the growing demand for more sustainable aquaculture practices. Several key innovations are shaping the future of fish nutrition.

5.1 Functional Feed Additives

Functional feed additives go beyond basic nutrition, providing additional benefits such as enhancing immunity, promoting gut health, and improving stress tolerance. These additives are increasingly being incorporated into aquaculture feeds to improve the overall performance and resilience of fish in intensive farming systems.

- **Prebiotics and Probiotics:** Prebiotics are non-digestible food ingredients that stimulate the growth of beneficial gut bacteria, while probiotics are live microorganisms that promote a healthy gut microbiota. Both play crucial roles in enhancing digestion, nutrient absorption, and immune function. For instance, probiotics such as *Lactobacillus* have been found to boost growth rates, improve feed conversion, and strengthen the immune system of fish. By enhancing gut health, these additives reduce the incidence of diseases and promote more efficient nutrient uptake.
- **Phytogenic Additives:** Plant-based compounds, such as essential oils, spices, and herb extracts, are gaining attention in aquaculture for their health-promoting effects. Ingredients like ginger, turmeric, and garlic are known to improve fish digestion and boost immunity. Ginger, in particular, has been shown to increase digestive enzyme activity, thereby improving feed efficiency and growth. Additionally, phytogenics possess antioxidant and anti-inflammatory properties, which help fish combat stress and infections.
- **Immunostimulants:** Immunostimulants are compounds that enhance the immune response of fish, making them more resistant to diseases. Beta-glucans, derived from yeast and fungi, are one of the most commonly used immunostimulants in aquaculture. These compounds have been shown to activate immune cells, such as macrophages, which play a critical role in defending against infections. Incorporating immunostimulants in aquafeeds can reduce the reliance on antibiotics, promoting healthier and more sustainable aquaculture practices.

5.2 Sustainable Feed Ingredients

Sustainability is a major concern in aquaculture, particularly regarding the reliance on fishmeal and fish oil as primary ingredients. Recent innovations focus on alternative protein and lipid sources that are both environmentally friendly and nutritionally adequate for fish. Sustainability is a major focus in modern aquaculture, particularly concerning the use of fishmeal and fish oil, which is traditionally derived from wild fish stocks. To address the environmental and economic concerns associated with these ingredients, researchers and industry leaders are exploring alternative protein and lipid sources that can replace or reduce the reliance on fishmeal and fish oil.

- **Insect Meal:** Insect proteins are gaining traction as a sustainable alternative to fishmeal. Black soldier fly larvae, mealworms, and other insects provide a high-quality protein source that is rich in essential amino acids. Studies have shown that insect meal can support similar growth rates and feed conversion ratios (FCR) to fishmeal in several aquaculture species, including tilapia and salmon. Insects also have the advantage of being produced on organic waste, further contributing to the circular economy and reducing environmental impact.

- **Algal Oil:** Microalgae are a rich source of omega-3 fatty acids, particularly eicosapentaenoic acid (EPA) and docosahexaenoic acid (DHA), which are essential for fish health and human consumption. Algal oil offers a sustainable alternative to fish oil, which is traditionally obtained from small pelagic fish. By incorporating algal oil into aquaculture feeds, the industry can reduce its dependence on wild fish stocks, helping to alleviate overfishing pressures. Moreover, algal oil has been shown to provide comparable levels of essential fatty acids, supporting optimal growth, immune function, and reproductive health in farmed fish.

5.3 Precision Feeding and Automated Systems

Technological advancements have revolutionized feeding practices in aquaculture, leading to the development of precision feeding systems that optimize feed delivery and reduce waste. These systems allow for more targeted feeding strategies, improving feed efficiency and fish growth.

- **Automated Feeding Systems:** Automated systems use sensors, cameras, and computer algorithms to monitor fish behavior and environmental conditions in real-time. These systems can adjust feeding rates based on factors such as fish size, feeding behavior, and water quality, ensuring that fish receive the appropriate amount of feed at the right time. By reducing overfeeding and minimizing feed waste, automated feeding systems enhance feed efficiency, reduce production costs, and contribute to better water quality by limiting the accumulation of uneaten feed in the system.
- **Nutrigenomics:** Nutrigenomics is an emerging field that studies how diet influences gene expression in fish. By understanding the relationship between nutrition and gene activity, researchers can develop more personalized feeding strategies that optimize growth, health, and stress resilience. For example, specific nutrients or feed additives may activate genes related to growth hormone production, immune function, or metabolic efficiency. Nutrigenomics offers the potential to create customized diets that enhance fish performance at different life stages or under specific environmental conditions, leading to more efficient and sustainable aquaculture production.

Table: Nutritional Physiology of Fish: A Foundation for Innovation

Section	Key Points
Overview of Fish Nutrition	- Fish nutrition affects growth, health, and feed conversion efficiency. - Nutrient requirements vary by species and life stages.
Digestive System of Fish	- Fish have varying digestive systems based on feeding habits (carnivorous, omnivorous, herbivorous). - Includes organs like the mouth, esophagus, stomach, intestines, liver, and pancreas.
Nutrient Requirements in Fish	- **Proteins**: Essential for tissue growth and metabolic processes. - **Lipids**: Source of energy and essential fatty acids. - **Carbohydrates**: Varying levels of digestibility among fish. - **Vitamins & Minerals**: Vital for physiological functions like immune response, osmoregulation, and bone development.
Feeding Behavior and Energy Metabolism	- Feeding strategies: Continuous feeders (herbivores), intermittent feeders (carnivores), and periodic feeders (species dependent on environmental factors). - Feeding schedules can optimize energy utilization.
Innovative Approaches Based on Physiology	- **Functional Feed Additives**: Probiotics, prebiotics, phytogenic additives, and immunostimulants improve fish health and nutrition. - **Sustainable Ingredients**: Insect meal, algal oil, and plant-based proteins replace fishmeal and fish oil. - **Precision Feeding**: Automated systems and nutrigenomics for tailored feeding strategies.
Functional Feed Additives	- Prebiotics and Probiotics improve gut health. - Phytogenics (e.g., ginger) enhance immunity and growth - Immunostimulants like beta-glucans support disease resistance.
Sustainable Feed Ingredients	- Alternative protein sources include insect meal. - Algal oil provides omega-3 fatty acids, reducing the need for fish oil.
Precision Feeding & Automated Systems	- Automated feeders adjust based on fish activity. - Nutrigenomics explores how nutrition affects gene expression to improve feed formulation.
Conclusion	- Fish nutrition innovations are essential for enhancing growth, feed efficiency, and sustainability in aquaculture.

Conclusion

Understanding the nutritional physiology of fish is fundamental to designing innovative approaches that enhance aquaculture productivity and sustainability. By tailoring diets to the specific digestive capacities and nutritional requirements of different fish species, aquaculture can achieve optimal growth, health, and feed efficiency. Ongoing research and technological advancements in functional feed additives, sustainable ingredients, and precision feeding systems are poised to revolutionize fish nutrition, addressing the growing demand for sustainable and healthy aquaculture products.

References

Ahmed, I., & Khan, M.A. (2004). Dietary Vitamin C Requirement of Fingerling Indian Major Carp. Fish Physiology and Biochemistry, 30(3), 189-198.

Cho, C.Y., & Bureau, D.P. (1995). Determination of the energy requirements of fish with particular reference to salmonids. Journal of Applied Ichthyology, 11(3-4), 223-230.

Cowey, C.B., & Sargent, J.R. (1979). Nutrition. In W.S. Hoar, D.J. Randall, & J.R. Brett (Eds.), Fish Physiology (Vol. 8, pp. 1-69). Academic Press.

Dabrowski, K. (1992). Ascorbic Acid and Fish Reproduction: Aspects of Nutritional Requirements, Gametogenesis, and Hormone Levels. Aquaculture, 93(2), 125-137.

Darias, M.J., et al. (2015). Vitamin D3 and Dietary Phosphorus Deficiency in Fish. Comparative Biochemistry and Physiology Part B, 181, 50-57.

De Silva, S.S., & Anderson, T.A. (1995). Fish Nutrition in Aquaculture. Chapman & Hall.

El-Dakar, A.Y., et al. (2004). Vitamin E Supplementation in Fish Diets. Aquaculture Research, 35(12), 1193-1200.

Francis, G., Makkar, H.P.S., & Becker, K. (2001). Antinutritional factors present in plant-derived alternate fish feed ingredients and their effects in fish. Aquaculture, 199(3-4), 197-227.

Gatlin, D.M. (2002). The Role of Minerals in Fish Nutrition. North American Journal of Aquaculture, 64(1), 44-53.

Gatlin, D.M., Barrows, F.T., Brown, P., Dabrowski, K., Gaylord, T.G., Hardy, R.W., Herman, E., Hu, G., Krogdahl, Å., Nelson, R., & Rust, M. (2007). Expanding the utilization of sustainable plant products in aquafeeds: a review. Aquaculture Research, 38(6), 551-579.

Glencross, B.D., Booth, M., & Allan, G.L. (2007). A feed is only as good as its ingredients – a review of ingredient evaluation strategies for aquaculture feeds. Aquaculture Nutrition, 13(1), 19-34.

Halver, J.E., & Hardy, R.W. (2002). Fish Nutrition (3rd ed.). Academic Press, San Diego, CA.

Hardy, R.W. (1997). Requirements for Trace Elements in Fish Feeds. Aquaculture, 151(1-4), 25-32.

Jobling, M. (2012). Fish Feed Optimization in Aquaculture. John Wiley & Sons.

Kaushik, S.J., & Seiliez, I. (2010). Protein and amino acid nutrition and metabolism in fish: current knowledge and future needs. Aquaculture Research, 41(3), 322-332.

Krogdahl, Å., Hemre, G.-I., & Mommsen, T.P. (2005). Carbohydrates in fish nutrition: Digestion and absorption in postlarval stages. Aquaculture Nutrition, 11(2), 103-122.

Kumar, V., Makkar, H.P.S., & Becker, K. (2011). Detoxified Jatropha curcas kernel meal as a dietary protein source: growth performance, nutrient utilization, and digestive enzyme activities in common carp (Cyprinus carpio L.) fingerlings. Aquaculture Nutrition, 17(4), 313-326.

Lall, S.P. (2002). The Minerals. In Fish Nutrition (3rd ed.), edited by Halver, J.E., & Hardy, R.W., 259-308. Academic Press, San Diego, CA.

Li, P., & Gatlin, D.M. (2006). Nutritional Strategies to Improve Fish Health. Journal of Animal Science, 84(suppl_1), E187-E197.

Lin, Y.-H., & Shiau, S.-Y. (2005). Dietary Magnesium Requirement of Juvenile Grass Shrimp. Aquaculture, 243(1-4), 431-438.

Lundstedt, L.M., Melo, J.F.B., & Moraes, G. (2004). Digestive enzyme response of pacu Piaractus mesopotamicus Holmberg (Osteichthyes, Characidae) to dietary protein level. Aquaculture Research, 35(11), 1083-1089.

National Research Council (2011). Nutrient Requirements of Fish and Shrimp. The National Academies Press.

Naylor, R.L., Hardy, R.W., Bureau, D.P., Chiu, A., Elliott, M., Farrell, A.P., Forster, I., Gatlin, D.M., Goldburg, R.J., Hua, K., & Nichols, P.D. (2009). Feeding aquaculture in an era of finite resources. Proceedings of the National Academy of Sciences, 106(36), 15103-15110.

NRC (National Research Council). (2011). Nutrient Requirements of Fish and Shrimp. National Academies Press, Washington, DC.

Peres, H., Lim, C., & Klesius, P.H. (2004). Nutritional Value of Dietary Phosphorus in Fish Nutrition. Reviews in Fisheries Science, 12(2-3), 113-123.

Stevens, C.E., & Hume, I.D. (1995). Comparative Physiology of the Vertebrate Digestive System. Cambridge University Press.

Tacon, A.G.J., & Metian, M. (2008). Global overview on the use of fish meal and fish oil in industrially compounded aquafeeds: Trends and future prospects. Aquaculture, 285(1-4), 146-158.

Tibbetts, S.M., Milley, J.E., & Lall, S.P. (2006). Apparent digestibility of protein and energy from common feed ingredients in Atlantic cod (Gadus morhua) and haddock (Melanogrammus aeglefinus). Aquaculture, 261(4), 1314-1327.

Tocher, D.R. (2003). Metabolism and functions of lipids and fatty acids in teleost fish. Reviews in Fisheries Science, 11(2), 107-184.

Waagbø, R. (1994). The Impact of Nutritional Factors on Fish Immune Responses. Annual Review of Fish Diseases, 4, 243-257.

Watanabe, T. (1988). Nutrition and Growth in Fish. In Fish Physiology Vol. XI, edited by Hoar, W.S., Randall, D.J., & Brett, J.R., 21-59. Academic Press, New York.

Wilson, R.P. (1994). Utilization of Dietary Carbohydrate by Fish. Aquaculture, 124(1-4), 67-80.

Wilson, R.P. (2002). Amino Acids and Proteins. In J.E. Halver & R.W. Hardy (Eds.), Fish Nutrition (3rd ed., pp. 143-179). Academic Press.

3

Sustainable Feed Ingredients The Future of Aquafeeds

Introduction

The rapid expansion of global aquaculture, driven by the increasing demand for high-quality, protein-rich seafood, has necessitated significant advancements in feed production. Historically, fishmeal and fish oil—derived from wild-caught fish—have been the cornerstone ingredients in aquafeeds, providing essential proteins and lipids for the optimal growth, health, and productivity of farmed species. However, the over-reliance on these finite marine resources has raised a host of environmental, economic, and ethical challenges. The depletion of wild fish stocks due to overfishing, habitat degradation, and climate change threatens the sustainability of global fisheries. Moreover, the volatility in the cost and availability of fishmeal and fish oil has placed increasing economic strain on the aquaculture industry, which is projected to grow substantially in the coming decades. These concerns have spurred the search for alternative, sustainable feed ingredients that can reduce dependence on wild-caught fish while maintaining or even improving the nutritional value and health outcomes for farmed fish. This chapter explores the ongoing evolution of sustainable feed ingredients and their crucial role in shaping the future of aquaculture. It highlights innovative approaches, such as plant-based proteins, insect meals, and microbial sources of omega-3 fatty acids, which offer promising solutions to enhance the sustainability of aquafeeds. By addressing both the environmental impact and economic viability of aquaculture feed production, these alternatives represent a pivotal shift towards a more responsible, resilient, and efficient aquaculture industry poised to meet the growing global demand for seafood.

The Need for Sustainable Aquafeeds

The exponential growth of aquaculture over the past few decades has positioned it as the fastest-growing food production sector globally, surpassing wild-caught fisheries in its contribution to the world's seafood supply. As aquaculture expands to meet rising consumer demands, the industry's reliance on traditional feed ingredients, primarily fishmeal and fish oil derived from wild-caught fish, has come under increasing scrutiny. The use of wild fish in aquafeeds presents significant challenges on multiple fronts. First, overfishing is a major concern. Fishmeal and fish oil production is directly tied to wild capture fisheries, placing immense pressure on already vulnerable marine populations. Many of these fish stocks are either overexploited or on the brink of depletion. The heavy exploitation of forage fish, which are essential in maintaining marine food webs, can disrupt ecosystems and threaten species that rely on these small fish for survival, creating a ripple effect throughout the environment.

In addition to overfishing, there is the environmental impact of harvesting wild fish for aquafeeds. This process not only depletes target fish species but can also lead to unintended consequences for marine biodiversity, as ecosystems are intricately balanced. Removing large quantities of fish can alter predator-prey dynamics, disrupt breeding patterns, and ultimately reduce biodiversity, which is critical for healthy and resilient marine ecosystems. Economic concerns are also mounting as the demand for fishmeal and fish oil continues to rise alongside aquaculture production. As wild fish resources become scarcer, the costs associated with sourcing fishmeal and fish oil increase, leading to higher production costs for fish farmers. This economic volatility jeopardizes the long-term sustainability and profitability of aquaculture operations, especially in regions where small-scale fish farming is a crucial livelihood.

Moreover, there are ethical and food security issues surrounding the use of wild fish for aquafeeds. Many of the fish used in fishmeal and fish oil production, such as anchovies and sardines, are nutrient-dense and could instead be used directly for human consumption, especially in food-insecure regions. This raises ethical questions about the allocation of these resources and the impact on global food security, as feeding fish to farmed fish or livestock may exacerbate hunger and malnutrition in vulnerable populations.

To mitigate these challenges, the development of sustainable feed ingredients has become a critical focus in aquaculture research and innovation. Sustainable alternatives to fishmeal and fish oil, such as plant-based proteins, insect meals, algae, and microbial oils, offer promising solutions to reduce dependence on wild-caught fish. These alternatives not only address environmental and ethical concerns but also provide economic advantages by stabilizing feed costs and improving the long-term viability of aquaculture. Furthermore, many of these sustainable ingredients are designed to maintain or enhance the nutritional quality of aquafeeds, ensuring that farmed fish remain healthy and productive.

In conclusion, transitioning to sustainable aquafeeds is essential for the future of aquaculture. By reducing the industry's reliance on finite wild fish resources, promoting biodiversity, and improving economic resilience, sustainable feed ingredients can support the growth of aquaculture while safeguarding environmental and food security concerns.

Key Sustainable Feed Ingredients

As the aquaculture industry continues to expand, there is an increasing need to diversify feed ingredients to reduce dependency on fishmeal and fish oil. Sustainable feed ingredients offer not only environmental and economic benefits but also the potential to enhance the nutritional quality of aquafeeds. Below is a detailed exploration of key sustainable feed ingredients shaping the future of aquafeed formulations.

1. Plant-Based Proteins

Plant-based proteins are the most established alternative to fishmeal and have been widely adopted across aquaculture species due to their abundance, cost-effectiveness, and sustainable production. These protein sources, derived from crops such as soy, corn, wheat, and peas, provide viable solutions, but they also pose challenges, including nutrient deficiencies and the presence of anti-nutritional factors (ANFs).

- **Soy Protein:** Soy protein is one of the most commonly used plant-based protein sources in aquafeeds due to its high protein content and affordability. It comes in several forms, including soybean meal, soy protein concentrate (SPC), and soy protein isolate (SPI). SPC and SPI are preferred in aquafeeds as they undergo processes to remove a significant portion of ANFs, such as protease inhibitors and oligosaccharides. Despite these advantages, soy protein can be deficient in essential amino acids (EAAs) like methionine and lysine, which are critical for the growth of carnivorous fish. Moreover, soy-based feeds may cause enteritis (intestinal inflammation) in some fish species like Atlantic salmon and rainbow trout, necessitating dietary supplementation with specific amino acids and careful diet formulation.
- **Corn Gluten:** Corn gluten meal is a by-product of corn processing and is another plant-based protein source used in aquafeeds. It is particularly valuable in feeds for omnivorous and herbivorous fish species like tilapia and carp due to its high protein and energy content. However, corn gluten meal lacks several EAAs, and its digestibility varies depending on the fish species and processing methods. This variability in digestibility can impact feed conversion efficiency, requiring precise formulation to balance nutrient requirements.

- **Pea Protein:** Pea protein is gaining popularity as a sustainable alternative due to its digestibility, protein content, and allergen-free profile. Pea protein has a lower environmental footprint compared to traditional crops like soy and corn, as it requires less water, fertilizer, and pesticides to grow. However, like other plant-based proteins, pea protein is deficient in some EAAs, and it may need to be supplemented with methionine or lysine to meet the specific dietary needs of carnivorous fish species, such as salmon and sea bass. Research has shown that pea protein can be used in combination with other plant-based proteins to create a more balanced amino acid profile in aquafeeds.

2. Insect Meal

Insect meal has emerged as a sustainable, high-protein alternative to fishmeal. Insects are part of the natural diet of many fish species, making them a biologically appropriate option. Moreover, insect farming is resource-efficient, requiring minimal land, water, and feed inputs. Insects like black soldier fly larvae, mealworms, and crickets can be cultivated on organic waste streams, contributing to a circular economy by upcycling waste into valuable feed.

- **Black Soldier Fly (BSF) Larvae:** BSF larvae are one of the most promising insect species for aquafeeds due to their ability to convert organic waste into high-quality protein. They can be reared on food waste, manure, or agricultural by-products, making them highly sustainable. BSF larvae contain high levels of protein (up to 60% dry weight) and fat (up to 35% dry weight), providing essential nutrients for fish. Additionally, BSF larvae are rich in antimicrobial peptides and lauric acid, which may enhance the immune function of farmed fish, reducing disease outbreaks. Studies have demonstrated that BSF larvae meal can replace a significant portion of fishmeal in the diets of species such as tilapia, trout, and shrimp, without negatively affecting growth, feed conversion ratios (FCR), or overall health.
- **Mealworms:** Mealworms (*Tenebrio molitor*) offer another insect-based protein source that can be used in aquafeeds. They are rich in protein (up to 50% dry weight) and fat, with a balanced amino acid profile. Their digestibility is high, making them suitable for a variety of aquaculture species. Mealworm farming is highly resource-efficient, requiring minimal water, land, and feed inputs. Research has shown that mealworm meal can successfully replace fishmeal in the diets of species like rainbow trout and sea bream, improving growth performance and feed efficiency.

3. Microalgae and Macroalgae

Algae, both microalgae and macroalgae (seaweed), are considered one of the most promising sustainable alternatives to fishmeal and fish oil due to their rich content of essential fatty acids, protein, vitamins, and minerals. Algae can be cultivated in non-arable areas and do not require freshwater, making them highly sustainable.

- **Microalgae:** Microalgae, such as Nannochloropsis, Schizochytrium, and Chlorella, are rich in omega-3 fatty acids, specifically eicosapentaenoic acid (EPA) and docosahexaenoic acid (DHA), which are critical for fish growth, health, and human nutrition. Microalgal oils can directly replace fish oil in aquafeeds, providing comparable levels of EPA and DHA. Some microalgae species can also accumulate high levels of protein (up to 50% dry weight) and essential amino acids, making them a valuable component of aquafeeds. Furthermore, microalgae cultivation does not compete with food crops for arable land or freshwater, and it can be integrated with carbon capture systems to reduce greenhouse gas emissions.
- **Macroalgae (Seaweed):** Seaweeds, such as kelp, red algae, and green algae, are abundant in coastal regions and can be farmed without the use of freshwater, fertilizers, or pesticides. Seaweeds are rich in polysaccharides, vitamins, minerals, and bioactive compounds with antioxidant and anti-inflammatory properties. These compounds have been shown to improve immune function, gut health, and stress resilience in farmed fish. Additionally, seaweed can act as a functional feed ingredient, promoting fish health and reducing the need for antibiotics or other chemical interventions. Seaweed farming is also beneficial for the environment, as it can mitigate ocean acidification, sequester carbon, and provide habitat for marine life.

4. Fermented Ingredients

Fermentation technology can enhance the nutritional quality and digestibility of plant-based feed ingredients by breaking down anti-nutritional factors and improving amino acid profiles. Fermented ingredients, such as fermented soybean meal, have been shown to improve fish growth, feed efficiency, and gut health.

- **Single-Cell Proteins (SCP):** SCPs, derived from microorganisms such as bacteria, yeast, and fungi, are emerging as a sustainable and innovative protein source for aquafeeds. These microorganisms can be cultivated using industrial by-products or waste streams, reducing their environmental footprint. SCPs are rich in protein (up to 70% dry weight), essential amino acids, and other bioactive compounds, making them a promising alternative to fishmeal. Methanotrophs, a type of bacteria that utilize methane as an energy source, are particularly promising as they can convert methane emissions into protein-rich biomass. This process not only reduces greenhouse gas emissions but also provides a sustainable feed ingredient for aquaculture.

5. Synthetic Amino Acids

To address the deficiencies of essential amino acids (EAAs) in some plant-based and alternative protein sources, synthetic amino acids such as lysine, methionine, and threonine are commonly added to aquafeeds. These synthetic amino acids are chemically identical to those found in fishmeal and help ensure that fish receive the precise nutrients needed for optimal growth, health, and feed efficiency. By supplementing plant-based diets with synthetic EAAs, aquafeed manufacturers can reduce their reliance on fishmeal while maintaining the nutritional quality of the feed.

6. By-Products from Food and Agriculture

By-products from the food and agriculture industries provide a sustainable way to reduce waste and enhance the nutrient content of aquafeeds. Animal by-products such as poultry by-product meal, blood meal, and feather meal are rich in protein and can replace fishmeal in many aquafeed formulations. These by-products are often more affordable than fishmeal and contribute to a circular economy by repurposing waste streams. Similarly, agricultural by-products like rice bran, wheat bran, and distiller's grains can serve as carbohydrate and protein sources in aquafeeds, reducing feed costs and reliance on marine resources.

Sustainable feed ingredients offer a pathway to reducing the environmental and economic pressures associated with traditional fishmeal and fish oil-based aquafeeds. From plant-based proteins and insect meals to innovative alternatives like microalgae, single-cell proteins, and by-products from food and agriculture, the aquaculture industry is making significant strides toward creating nutritionally balanced, eco-friendly feed options. These ingredients not only promote the growth and health of farmed fish but also contribute to the long-term sustainability of global aquaculture. By adopting a diversified approach to feed formulation, the industry can continue to meet the growing demand for seafood while minimizing its environmental impact and supporting global food security.

Advantages and Challenges of Sustainable Feed Ingredients

The future of aquafeeds hinges on the development of sustainable feed ingredients that not only meet the nutritional needs of farmed fish but also address the environmental and economic concerns associated with conventional feeds. While many of these ingredients offer considerable advantages, their widespread adoption also presents several technical and market-related challenges that must be overcome. Below, we delve deeper into the specific advantages and challenges of using sustainable feed ingredients in aquaculture.

Advantages

1. **Environmental Sustainability:** The aquaculture industry is under increasing scrutiny to reduce its reliance on finite marine resources, especially as wild fish stocks continue to decline. Fishmeal and fish oil are traditionally derived from small pelagic fish species like anchovies and sardines, which are often overexploited. Sustainable feed ingredients like plant proteins, insect meal, and algae can significantly reduce pressure on these wild fisheries, helping to restore balance in marine ecosystems. For instance, insect farming, particularly black soldier fly larvae (BSFL), is a highly efficient system that converts organic waste into valuable protein and fat sources, essentially creating a circular economy. Algae, both microalgae and seaweed (macroalgae), can be cultivated using non-arable land and saltwater, minimizing the need for freshwater and agricultural land that could otherwise be used for human food production. As a result, these alternatives offer a viable path to reducing the overall ecological footprint of aquaculture operations.
2. **Resource Efficiency:** One of the standout advantages of many sustainable feed ingredients is their resource efficiency. For example, insects can be farmed on waste products such as food scraps or agricultural by-products, turning low-value inputs into high-quality protein and fat. Black soldier fly larvae (BSFL) are particularly noteworthy, as they can grow on organic waste, including animal manure, food waste, or by-products from other industries, creating a zero-waste system. This approach offers a dual benefit: it reduces the environmental impact of waste while simultaneously producing valuable feed ingredients. Microalgae are another example of resource-efficient production. They require minimal space and can be cultivated in saline or wastewater environments, thus avoiding competition for freshwater and agricultural land. Algae farming also capture carbon dioxide during photosynthesis, contributing to carbon sequestration and further improving environmental sustainability.

3. **Cost-Effectiveness:** As global demand for seafood continues to grow, so too does the demand for fishmeal and fish oil, driving up prices. This price volatility poses a significant challenge to aquaculture producers, who need consistent and affordable feed sources. Sustainable feed ingredients, particularly plant-based proteins like soy and insect meals, offer a more stable and often lower-cost alternative. For example, soy protein concentrate (SPC) has become widely used in aquafeeds due to its affordability and wide availability. Although it lacks certain essential amino acids, its cost-effectiveness makes it a staple ingredient. Similarly, the ability to produce insect meal using low-cost substrates like waste reduces input costs, offering a promising alternative to more expensive fish-based ingredients. As these sustainable feed technologies continue to scale, the production costs are expected to decrease further, making them even more competitive in the global market.
4. **Nutritional Benefits:** Beyond their environmental and economic advantages, many sustainable feed ingredients offer unique nutritional benefits that contribute to the health and performance of farmed fish. Insects, for example, are naturally rich in high-quality protein, essential fatty acids, and trace minerals. Black soldier fly larvae (BSFL) contain antimicrobial peptides that may enhance the immune response of farmed fish, potentially reducing the need for antibiotics. In addition, algae, especially microalgae like *Schizochytrium* and *Nannochloropsis*, are rich in long-chain omega-3 fatty acids, particularly EPA and DHA, which are crucial for the growth and health of fish as well as human nutrition. Omega-3 fatty acids are typically supplied by fish oil in aquafeeds, but algae-based oils provide a direct plant-based source of these critical nutrients. Seaweeds (macroalgae) also offer bioactive compounds such as antioxidants, polysaccharides, and vitamins, which can enhance fish immune function, gut health, and overall growth performance, making them valuable functional feed additives.

Challenges

1. **Nutritional Imbalance:** While plant-based proteins and alternative ingredients offer promising sustainability benefits, they often lack the complete nutritional profile required by many fish species, particularly carnivorous species like salmon, trout, and sea bass. These fish have high requirements for specific essential amino acids (EAAs) such as methionine, lysine, and tryptophan, which are often deficient in plant-based proteins like soy, pea, and corn gluten. This nutritional imbalance necessitates the supplementation of synthetic amino acids or other protein

sources to ensure optimal growth and health. However, supplementation can increase the complexity and cost of feed formulation. Furthermore, while synthetic amino acids are bioavailable, they may not be as efficiently utilized by fish as those naturally occurring in fishmeal, potentially leading to suboptimal performance if not properly balanced.

2. **Anti-Nutritional Factors (ANFs):** Many plant-based ingredients contain anti-nutritional factors (ANFs) that can interfere with nutrient absorption and digestion in fish. For example, soybeans contain trypsin inhibitors, phytates, saponins, and other compounds that reduce the digestibility of protein and hinder the bioavailability of essential minerals like phosphorus. These ANFs can lead to intestinal inflammation, impaired growth, and reduced feed efficiency. While processing methods such as heat treatment, fermentation, or enzymatic hydrolysis can reduce the levels of ANFs, these processes increase production costs and may not completely eliminate all anti-nutritional effects. Addressing these issues requires continued research into optimizing the processing of plant-based ingredients to enhance their digestibility and minimize their negative impacts on fish health.
3. **Scaling Production:** While sustainable feed ingredients like microalgae, insect meal, and single-cell proteins (SCPs) show considerable potential, scaling up their production to meet the global demand for aquafeeds presents significant challenges. For instance, large-scale algae cultivation requires substantial infrastructure investment, and the costs associated with harvesting and processing algae into usable feed ingredients are currently higher than traditional fishmeal and fish oil production. Insect farming, although resource-efficient, requires automated systems to breed, raise, and process insects at scale. These production systems are still in the early stages of development, and achieving economies of scale will be essential for making insect meal a viable alternative to fishmeal. Additionally, single-cell proteins derived from bacteria, yeast, or fungi offer an innovative solution but face challenges related to fermentation costs, substrate availability, and public acceptance.
4. **Consumer Acceptance:** The public perception of alternative feed ingredients plays a significant role in determining their market success. While consumers are increasingly aware of the environmental benefits of sustainable aquaculture, the idea of using insect meal, microalgae, or waste-derived products in fish feed may raise concerns about food safety and quality. In particular, the use of insects in aquafeeds, while natural in many fish diets, may be met with skepticism from consumers unfamiliar

with the practice. Overcoming these barriers requires transparency and education about the safety, sustainability, and nutritional benefits of these alternative feed ingredients. Studies show that as consumers become more informed about the environmental impacts of aquaculture and the role of sustainable feeds in reducing those impacts, their acceptance of alternative feed ingredients improves. Clear labeling and communication from producers, combined with regulatory oversight, will be key to building trust and acceptance among consumers.

The integration of sustainable feed ingredients into aquaculture is a crucial step toward reducing the industry's reliance on finite marine resources and addressing the environmental, economic, and ethical challenges posed by traditional aquafeeds. While there are clear advantages, including enhanced sustainability, resource efficiency, cost stability, and potential nutritional benefits, significant challenges remain, particularly in terms of nutritional imbalances, anti-nutritional factors, production scalability, and consumer acceptance. Overcoming these challenges will require continued investment in research and development, technological advancements in production systems, and a collaborative effort between industry stakeholders, policymakers, and consumers to support the transition toward more sustainable aquafeeds. The future of aquaculture will depend on the ability to successfully integrate these alternative feed sources while maintaining the health, growth, and quality of farmed fish.

Table: Key Sustainable Feed Ingredients in Aquaculture and Their Advantages and Challenges

Feed Ingredient	Description	Advantages	Challenges
Plant-Based Proteins	Protein sources derived from crops like soy, corn, wheat, and peas are widely used as alternatives to fishmeal.	- Cost-effective and widely available. - Sustainable production. - Can reduce dependence on fishmeal.	- Nutrient deficiencies in essential amino acids (EAAs). - Presence of anti-nutritional factors (ANFs) like protease inhibitors and saponins. - Potential for causing enteritis in some species (e.g., Atlantic salmon).

Insect Meal	High-protein alternative to fishmeal derived from insects such as black soldier fly larvae and mealworms.	- Rich in protein and essential nutrients. - Can be produced using organic waste. - Environmentally sustainable with low resource requirements. - Enhances immune function and reduces disease outbreaks.	- Scaling production to meet global demand. - Requires optimized breeding and processing technologies. - Digestibility and nutrient consistency need further research.
Microalgae and Macroalgae	Algae are nutrient-rich sources of omega-3 fatty acids, protein, vitamins, and minerals. They include microalgae (e.g., Nannochloropsis, Chlorella) and macroalgae (e.g., kelp, seaweed).	- Rich in omega-3 fatty acids (EPA, DHA). - Grows in non-arable areas with no need for freshwater. - Supports carbon capture and reduces greenhouse gas emissions. - Contains bioactive compounds for improved fish health.	- High production costs for large-scale cultivation. - Harvesting and processing infrastructure is limited. - Requires further research into digestibility and formulation.
Fermented Ingredients	Includes fermented plant-based ingredients or single-cell proteins derived from microorganisms such as yeast and bacteria.	- Improved digestibility by breaking down ANFs. - Rich in protein and essential amino acids. - Supports circular economy by using waste streams.	- High production costs. - Requires further optimization of fermentation processes. - Variability in nutritional profiles based on the microorganism used.
Synthetic Amino Acids	Chemically identical to naturally occurring amino acids, synthetic forms like methionine and lysine are used to supplement plant-based diets.	- Allows precise nutrient formulations. - Reduces the need for fishmeal. - Addresses deficiencies in plant-based proteins.	- Costly to manufacture and incorporate. - Requires careful balance in diets for optimal utilization.
By-Products from Food and Agriculture	Animal by-products (e.g., poultry meal, blood meal) and agricultural by-products (e.g., rice bran, wheat bran) are repurposed for aquafeeds.	- Cost-effective and reduces waste. - Rich in protein and can replace fishmeal. - Supports circular economy.	- Nutritional variability depending on the source. - Potential contamination and variability in feed quality.

Conclusion

The development of sustainable feed ingredients marks a pivotal shift in the aquaculture industry, addressing critical environmental, economic, and ethical challenges linked to the heavy reliance on fishmeal and fish oil derived from wild-caught fish. As aquaculture plays an increasingly important role in global seafood production, sustainable alternatives, such as plant-based proteins, insect meal, algae, and single-cell proteins, provide viable pathways to reduce the pressure on overfished marine populations, mitigate biodiversity loss, and curb the environmental footprint of aquafeeds. These alternatives not only help protect marine ecosystems but also offer the potential for improved feed efficiency, nutritional benefits, and cost stability, all of which are essential for the industry's long-term sustainability.

Despite these advantages, challenges remain in achieving full-scale adoption of sustainable feed ingredients. Nutritional imbalances in some plant-based proteins, the presence of anti-nutritional factors, and the scalability of production for novel ingredients like microalgae and insect meal are significant hurdles that must be addressed. Moreover, public perception and acceptance of these alternative ingredients, especially those derived from unconventional sources such as insects or waste, will require transparent communication and education about their safety, efficacy, and sustainability.

Continued research and innovation are essential to overcoming these challenges, with a focus on optimizing ingredient formulations, enhancing digestibility, and improving processing technologies. Collaboration across industries, from feed manufacturers and aquaculture producers to policymakers and consumers, will be key to fostering the widespread adoption of sustainable aquafeeds. By integrating sustainable feed ingredients into aquaculture diets, the industry can not only meet the rising global demand for seafood but also contribute to a more responsible and eco-friendly food system. The future of aquafeeds lies in the collective efforts to balance fish health and growth with the health of our planet, ensuring a resilient and sustainable aquaculture industry for generations to come.

References

Aksnes, A., et al. (2006). Plant proteins as alternative fishmeal in diets for cod (Gadus morhua L.): Effects on growth and protein retention. Aquaculture, 255(1-4), 183-193.

Belghit, I., et al. (2019). Black soldier fly larvae meal can replace fishmeal in diets of sea-water phase Atlantic salmon (Salmo salar). Aquaculture, 503, 609-619.

Dalei, A., & Gouda, P. K. (2022). The role of microalgae in the production of aquafeeds for sustainable aquaculture. Aquaculture and Fisheries, 7(5), 257-268.

FAO (2020). The State of World Fisheries and Aquaculture 2020. Sustainability in Action. Food and Agriculture Organization of the United Nations.

Gasco, L., Henry, M., Piccolo, G., & Marono, S. (2016). The contribution of insect meal in fish nutrition. In Insects as food and feed: From production to consumption (pp. 129-147). Wageningen Academic Publishers.

Gatlin, D. M., et al. (2007). Expanding the utilization of sustainable plant products in aquafeeds: A review. Aquaculture Research, 38(6), 551-579.

Glencross, B. D. (2009). Exploring the nutritional demand for essential fatty acids by aquaculture species. Reviews in Aquaculture, 1(2), 71-124.

Glencross, B. D., et al. (2007). A feed is only as good as its ingredients–a review of ingredient evaluation strategies for aquaculture feeds. Aquaculture Nutrition, 13(2), 171-186.

Hardy, R. W. (2010). Utilization of plant proteins in fish diets: Effects of global demand and supplies of fishmeal. Aquaculture Research, 41(5), 770-776.

Hasan, M. R., & New, M. B. (2013). On-farm feeding and feed management in aquaculture. FAO Fisheries and Aquaculture Technical Paper No. 583. FAO, Rome.

Henry, M. A., Gasco, L., Chatzifotis, S., & Piccolo, G. (2015). Insect meals in fish nutrition. Reviews in Fisheries Science & Aquaculture, 23(4), 342-352.

Henry, M., Gasco, L., Piccolo, G., & Fountoulaki, E. (2015). Review on the use of insects in the diet of farmed fish: Past and future. Animal Feed Science and Technology, 203, 1-22.

Makkar, H. P. S., Tran, G., Heuzé, V., & Ankers, P. (2014). State-of-the-art on use of insects as animal feed. Animal Feed Science and Technology, 197, 1-33.

Naylor, R. L., et al. (2009). Feeding aquaculture in an era of finite resources. Proceedings of the National Academy of Sciences, 106(36), 15103-15110.

Naylor, R. L., et al. (2021). A 20-year retrospective review of global aquaculture. Nature, 591(7851), 551-563.

Olsen, R. L., & Hasan, M. R. (2012). A limited supply of fishmeal: Impact on future increases in global aquaculture production. Trends in Food Science & Technology, 27(2), 120-128.

Ritchie, H., & Roser, M. (2020). Environmental impacts of food production. Our World in Data.

Rust, M. B., et al. (2011). The future of aquafeeds: Proteins from new and unconventional sources. The Annual Review of Animal Biosciences, 1, 325-346.

Sealey, W. M., et al. (2011). Evaluation of algae as a potential source of dietary essential fatty acids for fish. Aquaculture, 316(1-4), 45-51.

Sørensen, M., et al. (2011). Alternative dietary protein sources for fishmeal in diet formulations for juvenile Atlantic cod (Gadus morhua). Aquaculture Research, 42(2), 291-302.

Tacon, A. G. J., & Metian, M. (2008). Global overview on the use of fishmeal and fish oil in industrially compounded aquafeeds: Trends and future prospects. Aquaculture, 285(1-4), 146-158.

Tibbetts, S. M., Milley, J. E., & Lall, S. P. (2011). Nutritional quality of seaweed products as a dietary protein source for Atlantic cod (Gadus morhua). Journal of Applied Phycology, 23(3), 401-408.

Turchini, G. M., et al. (2010). Fish oil replacement in finfish nutrition. Reviews in Aquaculture, 2(1), 10-57.

Van Huis, A. (2013). Potential of insects as food and feed in assuring food security. Annual Review of Entomology, 58, 563-583.

4

Functional Feed Additives for Enhanced Fish Growth and Health

1. Introduction

Functional feed additives are specialized, non-nutritive substances added to animal feed with the primary aim of enhancing health, growth, and performance beyond the basic nutritional requirements. Unlike traditional feed ingredients that primarily provide essential nutrients, functional additives are intended to optimize physiological functions and overall well-being in animals. In the context of aquaculture, these additives are particularly crucial for improving the efficacy of fish diets, as they can significantly influence various health and performance parameters.

Functional feed additives can be derived from a diverse array of sources, including:

- **Microorganisms**: Probiotics and prebiotics that support gut health and improve digestion.
- **Plants**: Phytochemicals, such as flavonoids and alkaloids, which have antioxidant, anti-inflammatory, and immunostimulatory properties.
- **Natural Compounds**: Various extracts and essential oils known for their therapeutic effects on fish health.

By integrating these additives into aquafeeds, aquaculture practitioners aim to promote fish welfare, enhance production efficiency, and ultimately contribute to sustainable fish farming practices.

Importance in Aquaculture

As global populations continue to rise, the demand for high-quality seafood has surged, leading to significant growth in the aquaculture sector. According to the Food and Agriculture Organization (FAO), aquaculture now contributes over 50% of the fish consumed worldwide. This rapid expansion of the industry brings with it numerous challenges, including:

- **Disease Outbreaks**: Fish farms are often susceptible to various pathogens, leading to outbreaks that can decimate stocks and harm the overall productivity of the aquaculture sector.

- **Fluctuating Feed Costs**: The prices of traditional feed ingredients are subject to market volatility, impacting the economic viability of fish farming operations.
- **Environmental Concerns**: The environmental footprint of aquaculture, including issues related to water quality, waste management, and habitat degradation, has become a growing concern.

In this context, functional feed additives have emerged as a viable solution to address these challenges. By optimizing fish growth, enhancing disease resistance, improving feed efficiency, and promoting overall health, these additives are now indispensable components of modern aquaculture practices.

Overview of Benefits

The benefits of functional feed additives in aquaculture are multi-faceted and encompass several key areas:

Enhanced Growth Performance

Functional additives are known to improve feed conversion ratios (FCR), which is a critical measure of growth efficiency in fish. By enhancing the digestibility and absorption of nutrients, these additives can promote faster growth rates, enabling fish to reach market size more quickly. Research indicates that the inclusion of probiotics and prebiotics in fish diets can lead to significantly improved growth performance, reducing the time and resources required for production.

Improved Health and Immunity

One of the most significant advantages of functional feed additives is their ability to bolster the immune system of fish. Additives such as immunostimulants enhance the production of immune cells and promote a robust immune response, reducing the incidence of disease. Moreover, functional additives can help mitigate stress in fish, which is a common issue in aquaculture settings. By improving overall health and reducing disease susceptibility, these additives contribute to more resilient fish populations.

Optimized Nutrient Utilization

Functional feed additives play a crucial role in optimizing nutrient utilization in aquaculture diets. By improving the digestibility and bioavailability of nutrients, these additives ensure that fish can efficiently absorb essential vitamins, minerals, and macronutrients. This leads to better feed efficiency, allowing farmers to maximize the benefits of their feed formulations while minimizing waste.

Reduced Environmental Impact

The incorporation of functional feed additives can also contribute to reducing the environmental impact of aquaculture. Improved nutrient absorption and reduced waste production lower the levels of nitrogen and phosphorus released into the aquatic environment, which can help mitigate issues such as eutrophication. By promoting more sustainable feeding practices, functional additives can enhance the ecological balance of aquaculture systems.

Economic Viability

Finally, the economic viability of fish farming is significantly enhanced by the use of functional feed additives. Increased growth rates and reduced disease incidence lead to lower production costs, allowing fish farmers to achieve higher profitability. The long-term economic benefits associated with the use of these additives make them an attractive investment for aquaculture operators looking to improve their production efficiency.

2. Types of Functional Feed Additives

Functional feed additives can be classified into several categories, each serving distinct purposes in enhancing fish growth and health. This section will discuss five major types of functional feed additives: probiotics, prebiotics, immunostimulants, phytogenics, and omega-3 fatty acids.

2.1 Probiotics

Definition and Function

Probiotics are defined as live microorganisms that, when administered in adequate amounts, confer health benefits to the host. In aquaculture, probiotics are primarily used to improve gut health and nutrient absorption in fish. They play a crucial role in maintaining a balanced gut microbiota, which is essential for the overall health and well-being of aquatic animals. A healthy gut microbiome can significantly influence digestion, nutrient uptake, and immune function, making probiotics a valuable component of modern aquafeeds.

Mechanisms of Action

Probiotics exert their beneficial effects through various mechanisms, including:

- **Competitive Exclusion**: Probiotics compete with pathogenic bacteria for binding sites and nutrients in the gut, thereby inhibiting their growth. This competition helps maintain a healthy balance of gut flora, reducing the risk of infections.
- **Modulation of Immune Response**: Probiotics can stimulate the immune system, enhancing the fish's ability to resist infections. They activate

immune cells and promote the production of antimicrobial substances, providing a robust defense against pathogens.

- **Improvement of Digestive Processes**: Some probiotics produce enzymes that aid in the breakdown of feed components, enhancing nutrient availability. This enzymatic activity can lead to better nutrient absorption and utilization, which is crucial for optimal growth.

Benefits for Fish Health and Growth

Numerous studies have demonstrated the positive effects of probiotics on fish growth and health. For instance, a study by Irianto and Austin (2002) reported that the incorporation of probiotics in fish diets led to improved growth rates, enhanced feed conversion ratios, and reduced mortality rates. Similarly, research by Banjara et al. (2018) found that probiotics significantly improved the gut health of Nile tilapia, resulting in better growth performance. Overall, the use of probiotics in aquaculture can lead to healthier fish populations with enhanced resilience to environmental stressors and diseases.

2.2 Prebiotics

Prebiotics are non-digestible food components that stimulate the growth and activity of beneficial gut bacteria. Unlike probiotics, which are live microorganisms, prebiotics serve as substrates for probiotics, promoting a healthy gut micro biome. Common sources of prebiotics include dietary fibers such as inulin and fructooligosaccharides, which are found in various plant-based foods.

Interaction with Gut Microbiota

Prebiotics influence the composition and activity of gut microbiota by providing a food source for beneficial bacteria. This leads to an increase in the population of these bacteria, which can enhance gut health and improve nutrient absorption. A well-balanced gut microbiome is critical for fish, as it supports digestion, metabolism, and immune function.

Benefits for Digestion and Immune Response

Research has shown that prebiotics can improve fish growth, digestion, and immune responses. For example, a study by Ghosh et al. (2013) demonstrated that the addition of inulin, a common prebiotic, to the diets of catfish resulted in improved growth performance and enhanced immune responses. Prebiotics have also been shown to reduce the incidence of gastrointestinal diseases in fish by promoting the growth of beneficial bacteria that inhibit pathogens. The synergistic effect of prebiotics and probiotics can create a more favorable gut environment, leading to better overall health in aquatic animals.

2.3 Immunostimulants

Immunostimulants are compounds that enhance the immune response in fish, making them more resilient to diseases. These substances can be derived from natural sources, including plants and microorganisms, and are incorporated into fish diets to improve immune function. By strengthening the immune system, immunostimulants help fish cope with environmental stressors and resist infections.

Types of Immunostimulants

Immunostimulants can be categorized into various types, including:

- **Beta-Glucans**: Derived from yeast and fungi, beta-glucans are known to enhance innate immunity by activating macrophages and other immune cells. They play a vital role in the fish immune response, helping to recognize and eliminate pathogens.
- **Herbal Extracts**: Compounds like garlic extract, echinacea, and others have been shown to stimulate immune responses in fish. These herbal extracts often contain bioactive compounds with antioxidant, antimicrobial, and anti-inflammatory properties.
- **Vitamins and Minerals**: Certain vitamins (e.g., vitamin C, vitamin E) and minerals (e.g., zinc, selenium) are essential for supporting immune function. Supplementing fish diets with these nutrients can enhance overall health and improve resistance to diseases.

Effects on Fish Immunity and Disease Resistance

The use of immunostimulants in aquaculture has shown promising results in enhancing fish immunity and disease resistance. For instance, research by Li et al. (2016) demonstrated that dietary supplementation with beta-glucans improved the immune responses of grass carp, leading to increased resistance to diseases such as viral infections. By incorporating immunostimulants into fish diets, aquaculture practitioners can promote healthier fish populations that are better equipped to withstand disease outbreaks and environmental stress.

2.4 Phytogenics

Phytogenics are bioactive compounds derived from plants, including essential oils, herbs, and spices. These compounds possess various health-promoting properties and are increasingly used in aquaculture to improve fish health and performance. Common sources of phytogenics include thyme, oregano, garlic, and turmeric, each of which offers unique health benefits for fish.

Bioactive Compounds and Their Functions

Phytogenics contain a range of bioactive compounds, such as flavonoids, alkaloids, and terpenes, which exhibit antioxidant, antimicrobial, and anti-inflammatory properties. These compounds can enhance fish health by:

- **Reducing Oxidative Stress**: Phytogenics help neutralize free radicals and reduce oxidative damage, which can lead to improved health and performance.
- **Inhibiting the Growth of Harmful Pathogens**: The antimicrobial properties of phytogenics can help inhibit the growth of harmful bacteria and fungi, thereby reducing the risk of infections in fish.
- **Modulating Digestive Processes**: Phytogenics can stimulate digestive enzymes and enhance gut health, improving nutrient absorption and overall growth performance.

Applications in Aquaculture

The incorporation of phytogenics into aquafeeds has shown positive effects on fish growth and health. For example, a study by Khosravi et al. (2017) found that dietary supplementation with thyme essential oil improved growth performance and feed efficiency in rainbow trout. Other studies have reported enhanced survival rates, reduced stress responses, and improved feed utilization in fish supplemented with phytogenic compounds. The versatility and health benefits of phytogenics make them valuable additions to aquaculture diets.

2.5 Omega-3 Fatty Acids

Importance in Fish Diets

Omega-3 fatty acids, particularly eicosapentaenoic acid (EPA) and docosahexaenoic acid (DHA), are crucial for fish growth and health. These fatty acids play essential roles in cellular function, brain development, immune response, and overall well-being. Due to their critical functions, adequate intake of omega-3 fatty acids is vital for optimal growth and health in aquaculture species.

Sources and Supplementation

Traditionally, fish oil has been the primary source of omega-3 fatty acids in aquaculture. However, sustainability concerns regarding fishmeal and fish oil production have prompted the exploration of alternative sources, such as algal oils and by-products from fish processing. Algal oils, in particular, have gained attention as a sustainable source of omega-3 fatty acids, offering a viable option for aquaculture diets without depleting fish stocks. Supplementing fish diets with omega-3 fatty acids can lead to improved growth rates, enhanced immune function, and overall health.

Effects on Growth and Overall Health

Studies have consistently shown that adequate intake of omega-3 fatty acids is linked to improved growth performance and health in fish. For instance, a study by Gjedrem (2000) reported that dietary omega-3 supplementation in salmonids resulted in increased growth rates and enhanced immune responses. Furthermore, omega-3 fatty acids are known to improve the fatty acid profile of fish, contributing to the overall nutritional quality of the fish flesh and benefiting human consumers.

3. Mechanisms of Action

Functional feed additives enhance fish growth and health through several complex mechanisms. Understanding these mechanisms is crucial for optimizing their application in aquaculture, as they can significantly impact fish performance, immune function, and overall well-being.

3.1 Enhancing Nutrient Utilization

Functional feed additives enhance nutrient utilization through multiple pathways that improve the digestive efficiency and nutrient absorption of fish.

3.1.1 Support for a Healthy Gut Microbiome

- **Role of Probiotics**: Probiotics, which are live beneficial microorganisms, colonize the intestinal tract of fish and contribute to gut health by establishing a balanced microbiota. This balance prevents the overgrowth of pathogenic bacteria, which can disrupt digestion and lead to health issues. For example, probiotics such as *Lactobacillus* and *Bifidobacterium* have been shown to effectively outcompete harmful bacteria like *Vibrio* species, commonly associated with disease outbreaks in aquaculture.
- **Role of Prebiotics**: Prebiotics are non-digestible food ingredients that promote the growth and activity of beneficial gut bacteria. Common prebiotics, such as inulin and oligosaccharides, serve as substrates for probiotics. They stimulate the proliferation of beneficial bacteria, thereby enhancing the overall health of the gut microbiome. A well-balanced microbiome not only aids in digestion but also contributes to the regulation of immune responses and nutrient absorption.

3.1.2 Enzyme Production and Nutrient Breakdown

- **Endogenous Enzyme Production**: Some probiotics can enhance the production of digestive enzymes in the gut of fish. For instance, studies have shown that *Bacillus* spp. can secrete enzymes that break down proteins, carbohydrates, and fats more effectively. This results in a more efficient digestion process, making essential nutrients more bioavailable.

- **Microbial Enzymes**: Certain probiotic strains produce specific enzymes, such as proteases, which help in the hydrolysis of proteins into peptides and amino acids. Similarly, amylases break down starches into simpler sugars, while lipases facilitate the digestion of fats. The resulting increase in free amino acids, sugars, and fatty acids enhances the nutritional profile available to fish.

3.1.3 Bioavailability Enhancement

- **Phytogenics and Digestive Processes**: Phytogenic compounds, derived from plants, play a crucial role in modulating digestive processes. Many phytogenics stimulate the secretion of bile acids and pancreatic juices, which are essential for the emulsification and digestion of dietary fats. Compounds such as saponins and flavonoids have been reported to increase the surface area of feed particles, facilitating better nutrient access during digestion.
- **Enhancing Nutrient Solubility**: Phytogenics can also increase the solubility of nutrients in the gut, leading to improved absorption. For example, the presence of certain essential oils can enhance the release of digestive enzymes and increase the permeability of intestinal cells, allowing for greater nutrient uptake.

3.1.4 Improved Feed Conversion Ratio (FCR)

- **Optimizing Feed Inputs**: Enhanced nutrient utilization leads to improved feed conversion ratios (FCR), a critical parameter in aquaculture. A better FCR indicates that fish are converting feed into body mass more efficiently, thus requiring less feed to achieve targeted growth rates. This not only reduces feed costs for aquaculture producers but also contributes to environmental sustainability by minimizing feed waste and resource usage.

3.2 Interaction with Fish Physiology

Functional feed additives can influence various physiological processes in fish, leading to improvements in health and growth.

3.2.1 Immune System Activation

- **Immunostimulant Mechanisms**: Immunostimulants such as beta-glucans derived from yeast and fungi activate the immune system by enhancing the activity of immune cells, including macrophages and neutrophils. When fish consume these compounds, their immune cells become more responsive to pathogens, leading to a more robust defense against infections.

- **Cytokine Production**: The introduction of immunostimulants can lead to increased production of cytokines, which are signaling molecules that regulate immune responses. Elevated levels of pro-inflammatory cytokines can help in the quick recruitment of immune cells to sites of infection, thereby enhancing the fish's ability to combat disease.

3.2.2 Regulation of Inflammation

- **Role of Omega-3 Fatty Acids**: Omega-3 fatty acids, particularly EPA and DHA, play essential roles in managing inflammation in fish. These fatty acids are precursors to anti-inflammatory mediators known as resolvins and protectins. When fish are supplemented with omega-3 fatty acids, the production of these mediators can help resolve inflammation, leading to quicker recovery from injuries or infections.
- **Cell Membrane Integrity**: Omega-3 fatty acids contribute to the structural integrity of cell membranes, making them more fluid and functional. This fluidity is vital for proper cellular signaling and the overall health of tissues, contributing to better growth rates and resilience against environmental stressors.

3.2.3 Cellular Function and Growth

- **Cell Signaling and Gene Expression**: Omega-3 fatty acids influence various cellular signaling pathways that regulate growth and development. They can modulate gene expression related to growth factors and metabolic pathways, enhancing the overall growth potential of fish.
- **Role in Reproductive Health**: Adequate intake of omega-3 fatty acids is also crucial for the reproductive health of fish. Research has shown that these fatty acids support gonadal development and improve egg quality in female fish, which is essential for successful breeding and hatchery operations.

3.2.4 Hormonal Regulation

- **Endocrine Interactions**: Some functional feed additives can influence the endocrine system of fish, thereby affecting growth and metabolism. For example, certain phytogenic compounds may stimulate the release of growth hormones, leading to enhanced growth rates. Additionally, these additives can help regulate metabolic hormones that control energy utilization and fat deposition in fish.

3.3 Modulation of Gut Health

Maintaining gut health is essential for fish, as it directly influences their ability to digest food, absorb nutrients, and resist diseases. Functional feed additives support gut health through various mechanisms.

3.3.1 Maintenance of Balanced Microbiota

- **Prevention of Dysbiosis**: Functional feed additives such as probiotics and prebiotics help maintain a balanced gut microbiota, which is crucial for preventing dysbiosis. Dysbiosis, characterized by an imbalance in gut microbial communities, can lead to poor digestion, nutrient malabsorption, and increased susceptibility to diseases.
- **Enhanced Microbial Diversity**: By promoting the growth of beneficial bacteria, functional feed additives increase microbial diversity in the gut. A diverse gut microbiome is associated with improved resilience against pathogens and better overall health.

3.3.2 Prevention of Pathogen Colonization

- **Competitive Exclusion Mechanism**: Probiotics can prevent the colonization of harmful pathogens in the gut through a mechanism known as competitive exclusion. By occupying binding sites and competing for available nutrients, probiotics limit the resources available to pathogenic bacteria, thereby reducing the risk of infections.
- **Production of Antimicrobial Compounds**: Some probiotic strains produce antimicrobial substances, such as bacteriocins and organic acids, that inhibit the growth of harmful bacteria. This production further enhances the protective effects of probiotics in aquaculture.

3.3.3 Enhanced Nutrient Absorption

- **Increased Intestinal Surface Area**: A balanced gut microbiome supported by functional feed additives enhances the absorptive capacity of the intestinal lining. The presence of beneficial bacteria can stimulate the development of intestinal villi, which increases the surface area available for nutrient absorption.
- **Improved Mucosal Health**: The production of short-chain fatty acids (SCFAs) by beneficial gut bacteria serves to nourish intestinal epithelial cells. SCFAs enhance the integrity of the gut lining and promote healthy mucosal immune responses, which collectively contribute to improved nutrient absorption.

3.3.4 Immune Response Enhancement

- **Gut-Associated Lymphoid Tissue (GALT)**: The gut is home to a significant portion of the immune system, particularly the gut-associated lymphoid tissue (GALT). Functional feed additives, particularly prebiotics and probiotics, can enhance the activity and development of GALT, leading to improved immune responses.
- **Production of Immunoglobulins**: A healthy gut microbiome stimulates the production of immunoglobulins (Ig), which are antibodies that play a critical role in the immune response. Increased levels of Ig in the gut can provide additional protection against pathogens, enhancing disease resistance in fish.

3.3.5 Stress Reduction

- **Impact of Stress on Gut Health**: Stress can negatively impact gut health, leading to increased susceptibility to diseases and poor growth performance. Functional feed additives, particularly probiotics, have been shown to alleviate stress responses in fish.
- **Stabilization of Gut Microbiota**: By promoting a stable and healthy gut microbiome, functional feed additives can reduce the adverse effects of stress. Stress reduction enhances feed intake, growth rates, and overall health, leading to improved production outcomes in aquaculture.

4. Impact on Growth Performance

The impact of functional feed additives on fish growth performance is well-documented in scientific literature, highlighting their potential to enhance aquaculture productivity. Numerous studies have demonstrated that incorporating these additives into fish diets can lead to significant improvements in growth rates, feed efficiency, and overall health. This section explores the evidence supporting these claims, comparisons between traditional diets and those enriched with functional additives, and the long-term benefits for aquaculture productivity.

4.1 Evidence from Studies

4.1.1 Probiotics and Growth Performance

Research indicates that probiotics positively influence growth performance across various fish species. For instance, Khani et al. (2013) found that the inclusion of *Lactobacillus* probiotics in the diets of rainbow trout resulted in enhanced weight gain and improved feed conversion ratios (FCR). The study reported a significant increase in average daily weight gain, with fish receiving probiotics showing a 15% higher growth rate compared to the control group.

Similarly, a meta-analysis by Zakeri et al. (2020) reviewed multiple studies on prebiotic supplementation in tilapia. The analysis concluded that diets enriched with prebiotics led to an average increase of 10% in growth performance, as evidenced by higher final weights and reduced feed wastage. These findings suggest that probiotics not only improve gut health but also enhance nutrient utilization, directly contributing to better growth outcomes.

4.1.2 Prebiotics and Their Effects

Incorporating prebiotics like inulin and fructooligosaccharides has been linked to improved growth metrics in various species. For instance, a study conducted by Ghosh et al. (2013) demonstrated that the addition of inulin to the diets of catfish resulted in a notable increase in weight gain and enhanced feed efficiency. The fish receiving prebiotic supplements exhibited a 20% improvement in growth performance compared to control groups. Furthermore, studies indicate that prebiotics can enhance the gut microbiota composition, resulting in improved digestive health and nutrient absorption. This is particularly beneficial in aquaculture, where diet plays a crucial role in the overall health and growth of fish.

4.1.3 Immunostimulants and Their Role in Growth

The inclusion of immunostimulants has also shown promising results in enhancing fish growth performance. Research conducted by Liu et al. (2016) revealed that fish fed diets containing beta-glucans, an effective immunostimulant derived from yeast, exhibited significantly higher growth rates compared to those fed standard diets. The study reported a 25% increase in body weight over the experimental period for fish receiving the immunostimulant. Additionally, the enhanced immune response resulting from immunostimulant supplementation leads to reduced disease incidence, further contributing to growth performance. Healthier fish require less energy for disease resistance, allowing for more efficient growth and feed utilization.

4.2 Comparisons between Traditional Diets and Those with Functional Additives

Comparative studies consistently illustrate that fish diets supplemented with functional feed additives outperform traditional diets in terms of growth performance and feed efficiency.

4.2.1 Traditional Diets versus Functional Additive Diets

A study by Alavi et al. (2016) compared the growth performance of juvenile seabass fed traditional diets against those supplemented with functional additives, including probiotics and phytogenics. The results showed that fish on the supplemented diets achieved 30% greater weight gain and significantly improved FCR compared to those on traditional diets. The enhanced growth was attributed to improved gut health and nutrient absorption facilitated by the functional additives.

4.2.2 The Role of Phytogenics

Research into the effects of phytogenics, such as essential oils and plant extracts, has further substantiated the benefits of functional feed additives. A study by Khosravi et al. (2017) demonstrated that the inclusion of thyme essential oil in the diet of rainbow trout improved growth performance by 18% compared to control diets. The researchers attributed this improvement to the antioxidant and antimicrobial properties of phytogenics, which promote better overall health and reduced stress in fish.

4.2.3 Meta-Analyses Supporting Findings

Several meta-analyses corroborate these findings by compiling data from multiple studies on the effects of functional feed additives. One notable meta-analysis conducted by Zakeri et al. (2020) reviewed the impact of various functional additives on growth performance across different fish species. The analysis confirmed that fish fed diets with functional additives consistently outperformed those on traditional diets, with average increases in weight gain ranging from 10% to 30%.

4.3 Long-Term Benefits for Aquaculture Productivity

The long-term benefits of using functional feed additives in aquaculture extend beyond immediate growth improvements. By enhancing growth rates and feed efficiency, these additives contribute to increased productivity and profitability for fish farmers.

4.3.1 Increased Profitability

Higher growth rates and improved feed efficiency translate into reduced production costs for aquaculture operations. With functional feed additives, farmers can achieve desired growth targets more quickly, allowing for shorter production cycles and greater turnover. This efficiency not only boosts profitability but also enables farmers to meet market demands more effectively.

4.3.2 Enhanced Disease Resistance

Healthier fish populations resulting from functional feed additive supplementation are more resilient to diseases. By improving immune responses and gut health, these additives reduce the incidence of infections and the need for costly treatments. This resilience leads to lower mortality rates and reduced reliance on antibiotics, promoting sustainable aquaculture practices.

4.3.3 Sustainability in Aquaculture

Functional feed additives contribute to more sustainable aquaculture by optimizing feed utilization and reducing waste. By improving FCR and promoting healthier fish populations, these additives minimize the environmental impact of aquaculture operations. Furthermore, healthier fish require fewer resources, reducing the strain on wild fish populations used for feed.

4.3.4 Regulatory and Market Trends

As the aquaculture industry faces increasing scrutiny over sustainability practices, the use of functional feed additives aligns with market trends and consumer preferences for healthier, responsibly produced seafood. By incorporating these additives, fish farmers can enhance the quality of their products and better position themselves in a competitive market.

5. Health Benefits

Functional feed additives significantly contribute to the health of farmed fish, affecting growth performance and overall well-being. The benefits extend to enhancing immune responses, reducing disease incidence, and improving fish welfare, ultimately fostering more sustainable aquaculture practices.

5.1 Enhanced Immune Responses

The ability of fish to resist diseases largely depends on their immune systems. Functional feed additives, particularly immunostimulants, play a pivotal role in boosting these immune responses.

5.1.1 Mechanisms of Immune Enhancement

- **Activation of Immune Pathways**: Immunostimulants, such as beta-glucans, polysaccharides, and certain peptides, act by triggering immune pathways. When introduced into the fish's diet, they stimulate toll-like receptors (TLRs) on immune cells, which lead to the activation of various signaling cascades that enhance the production and activity of immune cells.
- **Increase in Immune Cell Production**: Enhanced immune function is marked by increased levels of critical immune cells. For instance, the supplementation of fish diets with specific immunostimulants can lead to higher counts of leukocytes (white blood cells), such as macrophages and lymphocytes that are essential for recognizing and combating pathogens. A study by Lemme et al. (2009) reported significantly increased counts of phagocytic cells in fish fed diets enriched with immunostimulants.

- **Antibody Production**: Immunostimulants can also enhance the humoral immune response, leading to increased antibody production. For instance, fish fed with diets containing mannan-oligosaccharides exhibited higher antibody levels against specific pathogens, providing enhanced protection during disease challenges.

5.1.2 Evidence from Studies

Numerous studies have documented the effectiveness of immunostimulants in improving immune responses:

- **Enhancement in Resistance to Pathogens**: A trial involving European sea bass (*Dicentrarchus labrax*) showed that fish supplemented with beta-glucans had significantly lower mortality rates when exposed to *Vibrio anguillarum* compared to control groups. This illustrates the role of immunostimulants in enhancing disease resistance through improved immune responses.
- **Long-term Immune Benefits**: Research by Fadal et al. (2019) showed that fish receiving dietary supplementation with various herbal immunostimulants, including garlic and ginseng, had prolonged enhanced immune responses and reduced susceptibility to infections over time.

5.2 Reduction in Disease Incidence

5.2.1 Pathogen Resistance

The incorporation of functional feed additives in aquaculture diets significantly reduces the incidence of diseases among farmed fish. By enhancing immune function and supporting gut health, these additives play a crucial role in pathogen resistance.

- **Role of Probiotics in Disease Prevention**: Probiotics, such as Lactobacillus and Bacillus species, compete with pathogenic bacteria in the gut, effectively reducing their ability to colonize and cause infections. For example, a study by Adams et al. (2012) found that tilapia fed probiotics had significantly lower mortality rates during disease outbreaks caused by Aeromonas hydrophila, highlighting the direct impact of probiotics on disease resistance.
- **Strengthened Gut Barrier**: Functional feed additives can enhance the integrity of the gut barrier, reducing the entry of pathogens into the bloodstream. This is particularly important as a compromised gut can lead to increased susceptibility to diseases.

5.2.2 Evidence from Research

- **Case Studies**: In a study conducted by Sahu et al. (2013), the inclusion of prebiotics such as inulin in the diets of catfish resulted in not only improved growth performance but also reduced occurrences of enteric diseases. The study emphasized that the prebiotic effect supports beneficial gut microbiota, which in turn strengthens the fish's defense against pathogens.
- **Multi-Factorial Benefits**: Research indicates that the health benefits from functional feed additives can be multi-factorial. For instance, the use of a diet containing both probiotics and immunostimulants showed a synergistic effect, leading to improved health and lower disease incidences compared to diets containing only one type of additive.

5.3 Improvements in Overall Fish Welfare

5.3.1 Stress Reduction

Stress is a significant concern in aquaculture, as it can lead to reduced growth, poor health, and increased susceptibility to diseases. Functional feed additives contribute to fish welfare by alleviating stress levels through several mechanisms.

- **Enhancing Health and Resilience**: Healthier fish are inherently less susceptible to stressors. The presence of functional additives helps maintain a balanced gut microbiome, which is crucial for optimal digestion and nutrient absorption, reducing the stress associated with malnutrition or digestive disorders.
- **Reduced Cortisol Levels**: Stressful conditions, such as high stocking densities or poor water quality, can elevate cortisol levels in fish. Research by Klesius and Shoemaker (2005) found that fish receiving diets supplemented with functional additives exhibited lower cortisol levels, suggesting reduced physiological stress and improved welfare.

5.3.2 Behavioral Improvements

- **Natural Behavior Expression**: Healthy fish exhibit more natural behaviors, including foraging, social interaction, and territoriality. Functional feed additives promote better health, enabling fish to express these behaviors, which is critical for their overall well-being.
- **Aggression and Competition**: Healthy populations with good gut health and enhanced immune systems show reduced aggression and competition for resources, leading to a more stable social structure within the tank or pond.

5.3.3 Long-term Welfare Benefits

- **Sustainability and Resilience**: The long-term implementation of functional feed additives leads to sustained health benefits in fish. Fish that consistently receive these additives demonstrate not only improved growth rates but also increased survival rates, contributing to the overall sustainability of aquaculture systems.
- **Economic and Environmental Implications**: Healthy fish populations lead to higher yields and lower mortality rates, which are economically beneficial for fish farmers. Reduced disease incidence also translates to lower dependency on antibiotics and other treatments, thus promoting environmentally sustainable practices in aquaculture.

The health benefits associated with functional feed additives are critical to enhancing fish health, growth, and welfare in aquaculture. By improving immune responses, reducing disease incidences, and fostering overall fish welfare, these additives play a significant role in promoting sustainable practices. As aquaculture continues to expand in response to global seafood demands, the adoption of functional feed additives will be essential for ensuring the health of farmed fish populations and the viability of the industry.

6. Sustainability and Economic Aspects

The integration of functional feed additives in aquaculture is pivotal for fostering sustainable practices and enhancing the economic viability of fish farming operations. These additives contribute not only to improved fish health and growth but also to the overall sustainability of aquaculture systems, addressing environmental concerns while promoting economic benefits.

6.1 Role in Sustainable Aquaculture Practices

6.1.1 Enhancing Nutrient Utilization

Functional feed additives, such as probiotics and prebiotics, play a crucial role in enhancing the digestibility and bioavailability of nutrients in fish diets. By improving nutrient utilization, these additives help reduce feed waste and lower the overall environmental impact of aquaculture.

- **Improved Feed Efficiency**: By optimizing nutrient absorption, functional feed additives lead to improved feed conversion ratios (FCR). This means that fish can achieve their growth targets with less feed, thereby reducing the pressure on feed resources. A study by Gjedrem (2000) demonstrated that diets supplemented with omega-3 fatty acids not only improved fish health but also resulted in better feed efficiency.

- **Lowered Nutrient Excretion**: Enhanced nutrient utilization translates into lower levels of nutrient excretion into aquatic environments. This reduction in waste can mitigate the risk of water pollution and eutrophication, which are significant environmental concerns associated with intensive fish farming practices.

6.1.2 Minimizing Environmental Impact

Functional feed additives contribute to sustainability by addressing various environmental challenges:

- **Reduced Use of Antibiotics**: By improving fish health and resilience against diseases, functional additives can reduce the reliance on antibiotics and other chemical treatments. This is critical for maintaining ecosystem health and preventing the development of antibiotic-resistant pathogens.
- **Sustainable Feed Ingredients**: The exploration of alternative feed sources, such as plant-based ingredients and by-products, aligns with sustainability goals. Functional additives derived from sustainable sources can be integrated into fish diets, thereby reducing the reliance on traditional fishmeal and fish oil, which are often associated with overfishing.
- **Promoting Biodiversity**: Sustainable aquaculture practices, supported by functional feed additives, can enhance fish health and growth in a way that reduces pressure on wild fish populations. This promotes biodiversity in aquatic ecosystems while ensuring a stable supply of seafood.

6.2 Cost-Effectiveness of Using Functional Additives

The economic implications of incorporating functional feed additives into aquaculture are profound and multifaceted.

6.2.1 Improved Profitability

- **Enhanced Growth Performance**: Functional feed additives are associated with improved growth rates, which can significantly increase the yield of fish farms. The increase in biomass directly correlates with higher profits for fish farmers.
- **Reduced Treatment Costs**: Healthier fish populations require fewer medical treatments, translating into substantial cost savings. Research by Borski et al. (2009) indicated that the integration of probiotics into fish diets not only improved growth rates but also reduced treatment expenses related to disease outbreaks.

- **Long-term Economic Returns**: The initial investment in functional feed additives can yield long-term economic benefits. As farmers experience lower mortality rates, better growth performance, and reduced reliance on veterinary services, their overall operational costs decrease.

6.2.2 Market Competitiveness

- **Consumer Preferences**: As consumers increasingly seek sustainably sourced seafood, fish farmers incorporating functional feed additives may find a competitive advantage. Sustainable aquaculture practices enhance product quality and can lead to higher market prices.
- **Certification Opportunities**: Sustainable practices supported by functional feed additives can help farms obtain certifications (e.g., ASC, MSC) that appeal to environmentally conscious consumers. This can lead to better market access and premium pricing.

6.3 Future Trends and Research Directions

The future of functional feed additives in aquaculture appears promising, driven by ongoing research and technological innovations.

6.3.1 Innovations in Biotechnology

- **Biotechnology Advancements**: The exploration of biotechnological methods to develop novel functional additives is gaining momentum. Genetic engineering and synthetic biology could enable the production of tailored additives that optimize fish health and growth.
- **Microbiome Research**: Understanding the gut microbiome of fish is a burgeoning area of research. Insights into the interactions between functional additives and gut microbiota can lead to the development of more effective products that enhance fish health and nutrition.

6.3.2 Alternative Sources for Functional Additives

- **Exploration of Algae and By-products**: Algae, as a rich source of omega-3 fatty acids and other bioactive compounds, are increasingly being investigated as alternative sources for functional additives. The use of by-products from food processing also presents opportunities for sustainable ingredient sourcing, minimizing waste while enhancing fish diets.
- **Customized Functional Additives**: Future research may focus on developing customized functional additives that cater to the specific nutritional needs of different fish species, optimizing their growth and health under various farming conditions.

6.3.3 Regulatory and Market Trends

- **Regulatory Frameworks**: As the aquaculture industry evolves, regulatory frameworks surrounding the use of functional feed additives are likely to tighten. Research efforts will be essential to ensure that new additives meet safety and efficacy standards.
- **Consumer Education and Awareness**: Increasing consumer awareness regarding the benefits of sustainably farmed seafood will drive demand for products derived from aquaculture systems utilizing functional feed additives. Continued outreach and education will be crucial for promoting these practices.

The integration of functional feed additives in aquaculture systems is vital for achieving sustainability and economic viability. By enhancing fish health, improving feed efficiency, and minimizing environmental impacts, these additives provide numerous benefits for fish farmers. As research continues to uncover innovative solutions and alternatives, the role of functional feed additives will undoubtedly expand, paving the way for more sustainable and profitable aquaculture practices in the future.

7. Challenges and Considerations

The integration of functional feed additives in aquaculture presents various opportunities to enhance fish health and productivity. However, addressing the associated challenges and considerations is essential for optimizing their use and ensuring the long-term sustainability of aquaculture practices.

7.1 Potential Drawbacks and Limitations

7.1.1 Variability in Efficacy

Formulation and Composition:

- **Active Ingredients**: The effectiveness of functional feed additives is highly dependent on the concentration and type of active ingredients present. For instance, the viability and stability of probiotics can vary among different formulations, affecting their ability to colonize the gut and confer health benefits.
- **Processing Methods**: The processing methods used to produce additives, such as heat treatment or drying, can influence the bioactivity of compounds. For example, excessive heat can denature enzymes in phytogenic additives, diminishing their efficacy.

Species-Specific Responses

- **Tailored Approaches**: Fish species exhibit distinct physiological and dietary requirements. Additives that are beneficial for one species may not have the same effects on another due to differences in digestive systems, metabolic pathways, and immune responses. Research must focus on tailoring formulations to specific species for optimal results.

- **Growth Stages**: The growth stage of fish also plays a crucial role in their response to functional additives. For instance, juvenile fish may benefit more from certain probiotics that enhance growth, while adult fish may require different formulations to address health issues.

Environmental Conditions

- **Influence of Water Quality**: Parameters such as pH, salinity, and temperature can impact the bioavailability and effectiveness of functional feed additives. For example, probiotics may lose viability in poor water quality conditions, rendering them ineffective.
- **Stress Factors**: Environmental stressors, such as overcrowding or poor water quality, can also impair the benefits of functional additives. Stress can lead to immunosuppression, reducing the efficacy of immunostimulants and other health-promoting additives.

7.1.2 Interaction with Feed Components

Nutritional Interactions:

- **Inhibitory Effects**: Some functional additives may inhibit the absorption of essential nutrients. For example, high levels of certain phytogenics could potentially interfere with the bioavailability of proteins or minerals in fish diets, leading to suboptimal growth and health outcomes.
- **Digestibility Considerations**: The digestibility of feeds may be altered by the presence of certain additives. For instance, if probiotics enhance fiber degradation, they may inadvertently affect the digestibility of energy-rich components in the diet.

Synergistic Effects

- **Combination Strategies**: While some functional additives may work synergistically to enhance fish health, others could exhibit antagonistic effects. The challenge lies in understanding these interactions to formulate diets that leverage positive synergies and minimize negative impacts.
- **Trial and Error**: Fish farmers may need to conduct trials to identify optimal combinations of functional additives, leading to increased costs and resource usage during the formulation process.

7.2 Regulatory Aspects

7.2.1 Complex Regulatory Landscape

Variation by Region

- **Diverse Regulatory Frameworks**: The approval and use of functional feed additives are governed by different regulatory frameworks worldwide. For example, the European Union has strict regulations regarding the safety and efficacy of feed additives, while other regions may have more lenient requirements. This variability can complicate the international trade of aquaculture products.
- **Species-Specific Regulations**: Regulatory approvals may differ based on the specific fish species being targeted, as different species may have unique health concerns or dietary requirements.

Approval Processes

- **Lengthy Approval Times**: The approval process for functional feed additives can be lengthy, involving extensive pre-market testing to demonstrate safety, efficacy, and quality. This can delay the introduction of potentially beneficial additives to the market.
- **Post-Market Surveillance**: Regulatory agencies may require ongoing monitoring and reporting on the use and effects of approved additives, imposing additional burdens on fish farmers to comply with these regulations.

7.2.2 Compliance and Awareness

Staying Informed:

- **Evolving Regulations**: Regulations concerning functional feed additives are continually evolving as new research emerges. Fish farmers must remain informed about changes to ensure compliance and avoid legal issues.
- **Training and Education**: Ongoing education and training programs for fish farmers can enhance awareness of regulatory requirements and best practices for the use of functional additives in aquaculture.

Impact on Innovation

- **Hesitance to Innovate**: Strict regulatory frameworks may deter manufacturers from investing in research and development of new functional feed additives. This can hinder innovation and the introduction of novel solutions that could benefit the aquaculture industry.

- **Collaboration with Regulatory Bodies**: Collaborative efforts between researchers, industry stakeholders, and regulatory bodies can help streamline approval processes and encourage innovation while maintaining safety standards.

7.3 Need for Standardization in Product Formulation

7.3.1 Importance of Standardization

Consistency in Performance:

- **Batch Variability**: Variability in the quality and composition of functional feed additives can lead to inconsistent results in fish growth and health. Standardized formulations can help ensure that fish farmers receive reliable products that deliver consistent performance.
- **Efficacy Assessment**: Standardization allows for the establishment of benchmarks for evaluating the efficacy of functional additives, making it easier to compare products and make informed purchasing decisions.

Quality Control

- **Manufacturing Standards**: Implementing quality control measures during the manufacturing process can help minimize variability in product quality. This may involve standardized methods for sourcing raw materials, production processes, and final product testing.
- **Accreditation and Certification**: Developing accreditation programs for manufacturers of functional feed additives can help ensure that products meet established quality and safety standards.

7.3.2 Development of Industry Standards

Collaborative Efforts:

- **Industry Partnerships**: The establishment of industry standards for functional feed additives requires collaboration among researchers, manufacturers, regulatory agencies, and aquaculture practitioners. This collaborative approach can help ensure that standards are evidence-based and practical for real-world applications.
- **Conferences and Workshops**: Hosting conferences and workshops focused on functional feed additives can facilitate knowledge exchange and promote the development of best practices and standardization efforts.

Educational Initiatives:

- **Awareness Campaigns**: Raising awareness about the importance of standardized formulations and quality control can encourage manufacturers to adhere to best practices. Educational materials and resources can provide guidance on formulation strategies and quality assurance.

- **Training Programs**: Implementing training programs for fish farmers and industry professionals can enhance understanding of the significance of standardization and how to evaluate the quality of functional feed additives.

While functional feed additives present numerous advantages in aquaculture, addressing the associated challenges is vital for optimizing their use and ensuring sustainable practices. Understanding the potential drawbacks, navigating the complex regulatory landscape, and advocating for standardization in product formulation will empower fish farmers and stakeholders to maximize the benefits of these additives while minimizing risks. Continued research and collaboration within the aquaculture industry are essential for overcoming these challenges and fostering the development of innovative, effective, and sustainable solutions.

8. Conclusion

Functional feed additives are becoming increasingly vital in the aquaculture sector, where they play a critical role in enhancing fish growth, health, and overall productivity. The integration of these additives into aquaculture practices not only improves feed efficiency but also fosters better immune responses and reduces the incidence of diseases among farmed fish. As the demand for seafood rises globally, functional feed additives emerge as essential tools to ensure that aquaculture can sustainably meet these growing needs.

8.1 Significance of Functional Feed Additives

8.1.1 Enhancing Feed Efficiency

One of the primary advantages of functional feed additives is their ability to improve feed efficiency. By enhancing nutrient digestion and absorption, these additives optimize the utilization of feed, leading to improved growth rates and better overall performance in fish. This is particularly crucial in an era where feed costs represent a significant portion of aquaculture operating expenses. By increasing feed efficiency, farmers can achieve better economic returns while minimizing waste and environmental impact.

8.1.2 Supporting Immune Health

The immune health of fish is paramount for preventing disease outbreaks, which can devastate fish populations and impact aquaculture sustainability. Functional feed additives, particularly immunostimulants, activate immune pathways and boost the production of immune cells, making fish more resilient to infections. A healthier fish population not only leads to higher productivity but also enhances the overall welfare of the fish, ensuring ethical and sustainable farming practices.

8.1.3 Reducing Disease Incidence

The incorporation of functional feed additives is associated with a marked reduction in disease incidence among farmed fish. By promoting gut health and supporting immune function, these additives contribute to healthier aquatic ecosystems. This reduction in disease outbreaks can lead to significant savings for fish farmers by decreasing the need for costly treatments and minimizing losses due to mortality.

8.2 Meeting Global Seafood Demands

As the global population continues to grow, so does the demand for seafood. Aquaculture plays a pivotal role in meeting this demand, but it faces numerous challenges, including environmental sustainability, disease management, and feed resource limitations. The strategic use of functional feed additives can address these challenges by enhancing production efficiency and promoting the health of farmed fish. This not only supports the livelihoods of aquaculture producers but also contributes to food security at a global level.

8.3 Promoting Health and Sustainability

The commitment to sustainable aquaculture practices is becoming increasingly important in the face of environmental concerns. Functional feed additives can help minimize the environmental footprint of fish farming by improving nutrient utilization and reducing waste production. For instance, additives that enhance feed efficiency lead to lower feed input requirements, which in turn can lessen the impact on marine ecosystems and reduce reliance on wild-caught fish for feed.

Moreover, healthier fish populations supported by functional additives are better equipped to thrive in diverse environmental conditions, reducing the need for intensive management practices and chemicals. This aligns with the growing consumer demand for sustainably sourced seafood, making functional feed additives a critical component in the transition toward more responsible aquaculture.

8.4 Future Outlook

The future of functional feed additives in aquaculture is promising, driven by ongoing research and innovation in this field. As scientific understanding of the mechanisms underlying the effects of these additives deepens, researchers are likely to discover new compounds and formulations that can further enhance fish health and productivity.

Moreover, advancements in biotechnology, such as the development of novel probiotics, prebiotics, and phytogenic additives, hold great potential for improving the efficacy and specificity of functional feed additives. Continued collaboration between researchers, industry stakeholders, and regulatory agencies will be essential to bring these innovations to market while ensuring safety and efficacy.

8.5 Conclusion

In conclusion, the integration of functional feed additives into aquaculture practices represents a critical advancement in the industry. These additives not only enhance fish growth and health but also contribute to sustainable and economically viable fish farming. As the aquaculture sector evolves to meet the challenges of the future, the role of functional feed additives will become increasingly prominent, paving the way for healthier fish, sustainable practices, and a more resilient seafood supply chain. Through continued research, innovation, and commitment to best practices, the aquaculture industry can harness the full potential of functional feed additives to secure its future and address global seafood demands effectively.

References

Abdel-Tawwab, M., Samir, F., Abdel-Razek, N., & Sakr, S. F. (2020). Effects of dietary probiotic, Lactobacillus acidophilus, on growth performance, antioxidative status and immune response of Nile tilapia (Oreochromis niloticus) under sublethal malathion toxicity. Aquaculture, 529, 735674.

Adams, A., et al. (2012). Use of probiotics for disease prevention and enhanced immune response in tilapia. Aquaculture Research, 43(6), 827-838.

Ahmadifar, E., Fadaei Rayeni, M., Abbaszadeh, H., Dadar, M., Van Doan, H., & Dawood, M. A. (2021). The effects of dietary polyphenols on growth, immune response, oxidative stress, and disease resistance in aquatic animals: A review. Reviews in Aquaculture, 13(1), 181-199.

Alavi, S.E., et al. (2016). Comparative effects of functional feed additives versus traditional diets on growth performance and health in European seabass. Aquaculture Nutrition, 22(3), 546-555.

Balcazar, J.L., et al. (2006). Probiotics as biocontrol agents in aquaculture: A review of their role in fish health. Journal of Applied Microbiology, 100(1), 77-86.

Borski, R.J., et al. (2009). Economic and growth performance benefits of probiotics in aquaculture: Insights from a tilapia study. Aquaculture Economics & Management, 13(4), 234-248.

Dawood, M.A.O., & Koshio, S. (2016). Recent advances in the role of probiotics and prebiotics in fish nutrition and health. Reviews in Aquaculture, 8(1), 48-62.

Do Huu, H., et al. (2016). The role of dietary functional additives in enhancing fish immunity: A study in common carp. Fish & Shellfish Immunology, 55, 56-64.

Fadal, A., et al. (2019). Long-term benefits of herbal immunostimulants in boosting immune responses and disease resistance in farmed fish. Journal of Aquatic Animal Health, 31(2), 142-152.

Ghosh, S., Sinha, A., & Sahu, C. (2013). Dietary probiotics and prebiotics in fish and shellfish culture: a review. Aquaculture Research, 38(10), 1213-1232.

Gjedrem, T. (2000). Omega-3 fatty acids and their role in improving fish health and feed efficiency. Journal of Aquaculture, 183(3-4), 207-219.

Hoseinifar, S. H., Sun, Y. Z., Wang, A., & Zhou, Z. (2018). Probiotics as means of diseases control in aquaculture, a review of current knowledge and future perspectives. Frontiers in Microbiology, 9, 2429.

Irianto, A., & Austin, B. (2002). Use of probiotics to control furunculosis in rainbow trout, Oncorhynchus mykiss (Walbaum). Journal of Fish Diseases, 25(6), 333-342.

Khosravi, A., et al. (2017). Impact of thyme essential oil on growth, antioxidant capacity, and immune responses in rainbow trout. Fish & Shellfish Immunology, 67, 91-98.

Klesius, P.H., & Shoemaker, C.A. (2005). The impact of functional additives on stress responses and cortisol levels in fish. Journal of Fish Diseases, 28(7), 453-461.

Lemme, A., et al. (2009). The role of immunostimulants in enhancing the immune system of fish: A mechanistic approach. Fish Immunology Journal, 27(4), 350-365.

Liu, Y., et al. (2016). The effects of dietary beta-glucans on growth performance, immune response, and disease resistance in fish: A meta-analysis. Aquaculture, 452, 151-160.

Montalban-Arques, A., Flores-Félix, J. D., Gómez-López, L., Soria, A. G., & González-Andújar, J. L. (2021). Probiotics in aquaculture: A review of potential health and environmental benefits and risks. Science of the Total Environment, 753, 141968.

Newaj-Fyzul, A., Al-Harbi, A. H., & Austin, B. (2014). Review: Developments in the use of probiotics for disease control in aquaculture. Aquaculture, 431, 1-11.

Reverter, M., Bontemps, N., Lecchini, D., Banaigs, B., & Sasal, P. (2014). Use of plant extracts in fish aquaculture as an alternative to chemotherapy: Current status and future perspectives. Aquaculture, 433, 50-61.

Ringo, E., Olsen, R. E., Gifstad, T. O., Dalmo, R. A., Amlund, H., Hemre, G. I., & Bakke, A. M. (2010). Prebiotics in aquaculture: A review. Aquaculture Nutrition, 16(2), 117-136.

Rodriguez-Estrada, U., et al. (2013). Effect of mannan-oligosaccharides on growth performance and immune response in fish: A review. Fish & Shellfish Immunology, 34(6), 1493-1500.

Sahu, N.P., et al. (2013). Prebiotics as growth promoters and disease preventatives in aquaculture: A study in catfish. Aquaculture Nutrition, 19(1), 53-63.

Sealey, W. M., & Gatlin, D. M. (2001). Overview of nutritional strategies affecting the health of marine fish. Journal of Applied Aquaculture, 11(4), 49-61.

Silva, B. C., Vieira, F. N., Mouriño, J. L. P., Ferreira, G. S., & Seiffert, W. Q. (2016). Salts of organic acids improve the growth performance of juvenile Nile tilapia. Aquaculture International, 24(2), 731-744.

Situmorang, M.L., et al. (2020). Probiotics, prebiotics and synbiotics in fish nutrition: An overview of their role in immune enhancement. Aquaculture Research, 51(5), 1917-1931.

Soltanian, S., & Fereidouni, M. S. (2021). Application of probiotics and prebiotics in aquaculture: Review of recent developments and future perspectives. Reviews in Aquaculture, 13(2), 921-948.

Sørensen, M., et al. (2011). Evaluation of prebiotics and probiotics in fish nutrition and health: A case study of rainbow trout. Aquaculture Nutrition, 17(3), 481-494.

Turchini, G.M., et al. (2009). Functional feed additives and sustainable aquaculture: Advances and challenges. Reviews in Fisheries Science & Aquaculture, 17(4), 559-581.

Van Doan, H., Hoseinifar, S. H., Tapingkae, W., & Dawood, M. A. O. (2019). Effects of dietary olive and rosemary leaves extract on growth performance, antioxidant capacity, and immune response of Nile tilapia (Oreochromis niloticus) fingerlings. Fish Physiology and Biochemistry, 45(5), 1517-1526.

Wuertz, S., et al. (2012). Dietary supplementation of beta-glucans and their effects on fish immune systems. Journal of Aquatic Animal Health, 24(4), 211-218.

Zakeri, M., et al. (2020). Meta-analysis of the effects of functional additives on fish growth performance: A comprehensive review. Reviews in Aquaculture, 12(3), 1231-1251.

Zhou, Z., Ringø, E., Olsen, R. E., & Song, S. K. (2018). Dietary effects of soybean products on gut histology, gut microbiota, and digestive physiology of fish: A review. Aquaculture Nutrition, 24(6), 644-660.

5

Nutritional Genomics Tailoring Diets to Genetic Profiles

1. Introduction

Nutritional genomics, also known as nutrigenomics, is an emerging field that seeks to understand how individual genetic variations influence responses to specific nutrients and how diet, in turn, affects gene expression and overall health. By studying the complex interactions between nutrients and the human genome, nutritional genomics aims to develop personalized nutrition strategies that optimize health, prevent disease, and improve dietary outcomes for individuals based on their unique genetic makeup.

In the past, nutrition science largely followed a "one-size-fits-all" approach, with general dietary recommendations applied to broad populations. However, advances in genomics have shown that people respond differently to the same foods based on their genetic profile. Factors such as nutrient metabolism, absorption rates, and predispositions to certain diseases can vary significantly among individuals. These discoveries have led to a paradigm shift toward personalized nutrition, where diet plans are tailored to an individual's genetic predispositions, improving the effectiveness of dietary interventions and reducing disease risks.

Nutritional genomics spans several sub-disciplines, including nutrigenetics (the study of how genetic variations affect the body's response to nutrients) and nutrigenomics (how nutrients influence gene expression). These two aspects work together to help researchers understand the dual relationship between diet and genetics, contributing to the development of personalized diets that account for metabolic and physiological differences.

1.1 Overview of the Field's Scope and Importance

The growing interest in nutritional genomics reflects a global demand for precision nutrition to combat the rising burden of chronic diseases such as obesity, diabetes, cardiovascular disease, and cancer. Many of these conditions have genetic components, and their risk can be modulated through dietary choices. Thus, understanding the genetic factors that underlie nutrient metabolism is critical in crafting dietary strategies that can reduce susceptibility to such conditions and enhance overall well-being. For example, studies have shown that individuals with certain genetic variations may be more prone to developing elevated cholesterol levels when consuming saturated fats. Others might have altered folate metabolism due to polymorphisms in the MTHFR gene, which affects their risk of cardiovascular disease and certain types of cancer. Identifying these genetic differences can lead to specific dietary recommendations—such as lowering saturated fat intake or increasing folate consumption—to mitigate health risks.

In addition, nutritional epigenetics, a subset of nutritional genomics, focuses on how diet can induce epigenetic changes that regulate gene expression without altering the DNA sequence itself. Epigenetic modifications, such as DNA methylation or histone acetylation, can be influenced by certain nutrients, leading to changes in disease susceptibility, growth, and metabolism. This area of research shows that not only can genetics influence nutrition, but nutrition can also influence gene expression in a way that affects long-term health outcomes.

1.2 Historical Context and Evolution

The roots of nutritional genomics trace back to early discoveries in genetics and nutrition, particularly the recognition that certain genetic disorders, such as phenylketonuria (PKU), are directly influenced by dietary intake. Individuals with PKU, a rare genetic disorder, cannot metabolize the amino acid phenylalanine properly, leading to severe neurological damage unless they follow a strict low-phenylalanine diet. This finding laid the foundation for understanding how genetic differences could impact nutrient metabolism and set the stage for more complex interactions between diet and genetics. With the advent of the Human Genome Project in the early 2000s and subsequent advancements in genome-wide association studies (GWAS) and high-throughput sequencing technologies, scientists began to explore the broader implications of genetic variation in human nutrition. GWAS enabled researchers to identify common genetic variants associated with traits related to nutrition, such as lactose intolerance, gluten sensitivity, and vitamin metabolism. These developments sparked a surge of interest in how personalized dietary recommendations could be formulated based on genetic profiles.

As technological capabilities have improved, so too has the ability to integrate large-scale data sets from genomics, transcriptomics, metabolomics, and bioinformatics, further enhancing our understanding of nutritional genomics. Today, personalized nutrition is becoming more accessible to the public through direct-to-consumer genetic testing services that provide insights into individual nutritional needs based on genetic data. This has led to a broader movement toward precision health, where personalized diet and lifestyle recommendations aim to improve individual well-being.

1.3 Objectives and Purpose of Nutritional Genomics

The primary objective of nutritional genomics is to bridge the gap between genetic variability and dietary interventions by using genomic information to create customized nutritional strategies that promote optimal health and disease prevention. This approach is poised to revolutionize traditional dietary guidelines by accounting for individual differences in nutrient requirements, metabolism, and disease risk factors.

Key goals of nutritional genomics include:

1. **Personalized Nutrition Plans**: Developing tailored dietary recommendations based on an individual's genetic profile to optimize health outcomes and prevent chronic diseases.
2. **Disease Prevention**: Identifying genetic susceptibilities to conditions such as heart disease, diabetes, and cancer, and using diet as a preventive tool to mitigate these risks.
3. **Improved Nutrient Metabolism**: Understanding how genetic differences affect nutrient absorption, metabolism, and utilization, leading to more effective nutritional strategies.
4. **Epigenetic Modulation**: Investigating how dietary components influence gene expression and how these epigenetic changes can contribute to health and disease.
5. **Optimizing Public Health**: Enhancing public health guidelines by considering genetic diversity within populations and promoting more precise, effective dietary recommendations.

This chapter provides an overview of the fundamental principles of nutritional genomics, the methodologies used in this rapidly advancing field, and its potential applications in personalizing diets to improve health outcomes. Through case studies, emerging research, and examples of successful applications, the following sections will demonstrate how nutritional genomics is poised to transform healthcare and nutrition science in the 21st century.

1.4 Relevance to Contemporary Healthcare

In today's healthcare landscape, chronic diseases such as obesity, cardiovascular disease, and type 2 diabetes are at epidemic levels worldwide. A significant portion of these conditions can be attributed to poor diet, lifestyle choices, and, importantly, genetic predispositions. While public health initiatives often focus on improving dietary quality across populations, individual genetic variability means that a one-size-fits-all approach may not be effective in addressing these conditions. Nutritional genomics offers a solution to this issue by providing personalized recommendations that account for the genetic factors influencing disease susceptibility. By leveraging genomic information, healthcare providers can offer more precise dietary interventions that cater to the specific metabolic and physiological needs of their patients. This approach not only improves the efficacy of dietary recommendations but also empowers individuals to take a proactive role in managing their health.

Moreover, nutritional genomics holds great promise in precision medicine, where diet becomes an integral component of individualized treatment plans. For example, patients undergoing cancer treatment may receive personalized nutritional support based on their genetic makeup, enhancing recovery and treatment outcomes.

As we move toward a future of precision healthcare, nutritional genomics will play an increasingly vital role in optimizing diet, improving health outcomes, and preventing disease. The following chapters will explore in detail the methodologies and applications of this promising field, highlighting its potential to revolutionize both clinical nutrition and public health strategies.

2. The Basis of Nutritional Genomics

Nutritional genomics, also known as nutrigenomics, is a cutting-edge field that focuses on the relationship between an individual's genetic makeup and their response to specific nutrients. It delves into how genetic variations influence nutrient metabolism, absorption, and utilization, ultimately affecting health outcomes. This section explores the core principles of nutritional genomics, particularly the role of genetic variations such as single nucleotide polymorphisms (SNPs), and the emerging field of nutritional epigenetics.

2.1 Genetic Variations and Nutritional Response

One of the foundational concepts of nutritional genomics is understanding how genetic variations impact nutritional response. Single nucleotide polymorphisms (SNPs) are the most common type of genetic variation among individuals. SNPs represent changes in a single nucleotide—one of the building blocks of DNA—at a specific position in the genome. These variations can significantly affect how individual's process and respond to nutrients, which in turn can influence their overall health, disease risk, and response to dietary interventions.

Certain SNPs have been directly linked to nutrient metabolism and related health outcomes. For example:

- **FTO Gene and Obesity**: The **FTO (fat mass and obesity-associated) gene** has been identified as a key player in obesity and metabolic disorders. SNPs in the FTO gene have been shown to influence appetite regulation, energy expenditure, and fat accumulation. Individuals carrying certain variants of the FTO gene are more likely to experience difficulty in maintaining a healthy weight, even when consuming a similar diet as individuals without the variant. Understanding this genetic predisposition allows for the development of personalized dietary strategies to mitigate obesity risk.
- **MTHFR Gene and Folate Metabolism**: Another well-known example is the **MTHFR (methylenetetrahydrofolate reductase) gene**, which plays a critical role in folate metabolism and the conversion of homocysteine to methionine. SNPs in the MTHFR gene can reduce the efficiency of this process, leading to elevated levels of homocysteine, which is a risk factor for cardiovascular disease and certain types of cancer. Individuals with MTHFR polymorphisms may require higher intakes of folate or specific forms of folate supplementation (such as 5-methyl-tetrahydrofolate) to compensate for this genetic variation and reduce disease risk.
- **Lactase Persistence and Lactose Intolerance**: Another example of genetic variation influencing nutrition is lactase persistence, which determines whether individuals can digest lactose, the sugar found in milk, beyond infancy. In populations with high frequencies of lactase persistence, individuals can continue to consume dairy products without digestive issues. Conversely, in populations where lactase persistence is low, lactose intolerance is common, and individuals may need to avoid or limit dairy intake. This is a classic example of how understanding genetic variation can inform dietary recommendations at both individual and population levels.

These genetic differences highlight the importance of personalized nutrition, as people with different genetic profiles may require tailored dietary interventions to optimize their health outcomes. Nutrigenetics, a subfield of nutritional genomics, is focused on studying these genetic variations and their impact on nutrient requirements, metabolism, and overall health.

2.2 Nutritional Epigenetics

Nutritional epigenetics is an exciting and rapidly growing area of research that examines how nutrients and bioactive food components can influence gene expression through epigenetic mechanisms. Unlike genetic variations, epigenetic changes do not involve alterations to the DNA sequence itself. Instead, they affect the way genes are expressed or silenced through modifications to DNA or associated proteins, such as DNA methylation, histone modification, and non-coding RNAs. These changes can have long-term effects on an individual's health and can even be passed down to future generations.

- **DNA Methylation**: One of the most studied epigenetic modifications is **DNA methylation**, which involves the addition of a methyl group to cytosine bases in the DNA molecule. Methylation typically suppresses gene expression by preventing transcription factors from binding to DNA. Nutrients such as **folate, vitamins B6** and **B12**, and **methionine** are involved in one-carbon metabolism, which provides methyl groups for DNA methylation. Diets rich in these nutrients can positively influence DNA methylation patterns, potentially reducing the risk of conditions such as cardiovascular disease and certain cancers. For example, folate deficiency has been linked to aberrant DNA methylation patterns that may contribute to tumor development.
- **Histone Modifications**: Another key epigenetic mechanism is the modification of histones, the proteins around which DNA is wrapped. These modifications, such as **histone acetylation** and **histone methylation**, can either promote or inhibit gene expression depending on the nature of the modification. Certain nutrients, such as **polyphenols** found in fruits and vegetables, have been shown to influence histone modifications. For example, **resveratrol**, a polyphenol in grapes, has been linked to beneficial histone modifications that may protect against inflammation and oxidative stress.

Research in nutritional epigenetics has demonstrated that diet can have a significant impact on gene expression, with implications for the prevention and management of chronic diseases. For instance, a diet rich in fruits, vegetables, and whole grains has been shown to positively influence the expression of genes related to inflammation and oxidative stress, which are key factors in the development of cardiovascular disease, diabetes, and cancer. Conversely, diets high in processed foods, saturated fats, and sugars can lead to detrimental epigenetic changes that increase the risk of these diseases.

One of the most compelling aspects of nutritional epigenetics is the concept that epigenetic modifications are reversible. This means that by making positive dietary changes, individuals can potentially reverse harmful epigenetic marks and improve their health outcomes. For example, studies have shown that weight loss through a healthy diet and lifestyle can reverse DNA methylation patterns associated with obesity and metabolic disorders, leading to improved metabolic health. Nutritional epigenetics also has important implications for fetal programming and developmental origins of health and disease (DOHaD). The in utero environment, influenced by maternal diet, can lead to epigenetic modifications that affect the health of the offspring throughout their lifetime. For instance, maternal folate intake during pregnancy is critical for proper DNA methylation, which is essential for the development of the fetal nervous system. Insufficient folate intake can lead to neural tube defects and other developmental abnormalities. This highlights the importance of optimal nutrition not only for the individual but also for future generations.

In summary, nutritional epigenetics provides a framework for understanding how diet influences gene expression and how these changes can have profound effects on health and disease risk. By targeting epigenetic mechanisms through dietary interventions, it may be possible to prevent or manage chronic diseases and improve health outcomes across generations. This emerging field is poised to play a crucial role in the future of personalized nutrition and precision medicine, where dietary recommendations are tailored to an individual's genetic and epigenetic profile.

As the field of nutritional genomics continues to evolve, it holds great potential for revolutionizing dietary guidelines, offering more personalized and effective nutritional strategies to improve health outcomes and reduce disease risk across populations.

3. Methodologies in Nutritional Genomics

Nutritional genomics employs a variety of advanced methodologies to explore the intricate relationships between diet, genetics, and health. These techniques enable scientists to investigate how individual genetic differences influence responses to dietary components, shedding light on the mechanisms that drive nutrient metabolism and health outcomes. The integration of genomics, transcriptomics, metabolomics, and bioinformatics provides a comprehensive understanding of how diet interacts with an individual's genetic profile. Below are the key methodologies utilized in nutritional genomics:

3.1 Genome-Wide Association Studies (GWAS)

Genome-Wide Association Studies (GWAS) are a fundamental tool in nutritional genomics, as they allow for the identification of genetic variants associated with specific health traits or diseases. By examining large datasets of individuals' genomes, GWAS identify single nucleotide polymorphisms (SNPs) that are linked to nutrient metabolism, food intolerances, and diet-related diseases.

This methodology provides insights into how genetic variations contribute to differential responses to diet and nutritional interventions. For example, GWAS studies have revealed SNPs associated with lactose intolerance, helping to predict an individual's ability to digest lactose-containing foods. This knowledge is crucial in developing dietary recommendations for people with genetic predispositions, such as those who are genetically unable to produce sufficient lactase, the enzyme needed to digest lactose. Another example is the identification of SNPs in genes like APOA5, which are linked to triglyceride levels, helping guide dietary advice related to fat intake. GWAS findings are instrumental in creating personalized nutrition plans, where specific genetic variations can guide dietary choices to mitigate disease risk or optimize health outcomes. They also offer a pathway for future research into new biomarkers that could predict individual responses to different diets.

3.2 Transcriptomics

Transcriptomics focuses on analyzing RNA transcripts to understand how gene expression changes in response to dietary components. By studying RNA levels, scientists can track how certain foods or nutrients activate or suppress the expression of particular genes, offering insights into the biological pathways involved in nutrient metabolism and health. Transcriptomic studies utilize techniques such as RNA sequencing (RNA-seq) to measure the abundance of RNA molecules, revealing which genes are active at a given time. This data is valuable for understanding how specific nutrients modulate gene expression. For example, research has demonstrated that omega-3 fatty acids can upregulate genes involved in anti-inflammatory pathways while downregulating genes associated with pro-inflammatory responses. These gene expression changes provide a molecular basis for the health benefits of omega-3-rich diets, including their protective effects against cardiovascular diseases and chronic inflammation. Transcriptomics is essential for identifying diet-gene interactions and offers a mechanistic understanding of how individual dietary choices can influence gene expression patterns, ultimately affecting health outcomes. These insights could lead to the development of tailored nutritional interventions to prevent or manage diseases.

3.3 Metabolomics

Metabolomics is the comprehensive study of the small molecules—known as metabolites—produced during metabolic processes. This approach provides a snapshot of the biochemical activities within an organism at a given time, reflecting the physiological responses to dietary intake. Metabolomics enables the identification of metabolic pathways that are influenced by both genetic variations and dietary components. Metabolomics analyses typically involve sampling biological fluids such as blood, urine, or tissue extracts, and using techniques like mass spectrometry (MS) and nuclear magnetic resonance (NMR) to identify and quantify metabolites. These metabolites can serve as biomarkers, reflecting an individual's metabolic state, nutrient deficiencies, or risk of diseases like cardiovascular disease, diabetes, or obesity. For instance, specific metabolites associated with fatty acid metabolism may indicate an increased risk for cardiovascular disease, offering a predictive tool for early dietary intervention. Metabolomics is pivotal in personalized nutrition, as it links diet to metabolic responses and helps identify how genetic variants influence nutrient utilization. By tailoring dietary recommendations to optimize metabolic health based on an individual's metabolomic profile, this approach enables more precise and effective nutritional interventions.

3.4 Bioinformatics

The vast amount of data generated from nutritional genomics research requires sophisticated computational methods for analysis. Bioinformatics plays a critical role in managing and interpreting this complex data, integrating findings from GWAS, transcriptomics, and metabolomics to identify meaningful patterns and relationships between diet and genetic profiles. Bioinformatics tools utilize algorithms, machine learning, and data visualization techniques to process large datasets and uncover associations between genetic variants and nutrient-related phenotypes. These methods help researchers identify biomarkers, predict dietary responses, and design personalized nutrition plans that cater to an individual's unique genetic and metabolic profile. For example, bioinformatics has been used to create nutritional recommendation systems that integrate an individual's genetic data, food preferences, and health goals to generate tailored dietary advice. By employing models that predict nutrient-gene interactions, bioinformatics facilitates the development of personalized interventions to improve health outcomes and prevent disease.

Bioinformatics also aids in identifying gene-nutrient interactions and deciphering how various components of the diet influence molecular pathways across different populations. This technology-driven approach ensures that nutritional genomics can achieve its full potential in enhancing human health through precision nutrition.

The methodologies used in nutritional genomics are transforming our understanding of the complex interactions between diet, genetics, and health. By integrating insights from GWAS, transcriptomics, metabolomics, and bioinformatics, researchers are uncovering the molecular basis of individual responses to nutrients, paving the way for personalized nutrition. These cutting-edge approaches provide a framework for optimizing dietary strategies based on genetic profiles, offering a promising future for precision health and disease prevention.

Table: Key Aspects and Methodologies in Nutritional Genomics

Section	Description
The Basis of Nutritional Genomics	Nutritional genomics (nutrigenomics) focuses on the relationship between an individual's genetic makeup and their response to specific nutrients. It explores how genetic variations influence nutrient metabolism, absorption, and utilization, impacting health outcomes.
Genetic Variations and Nutritional Response	Genetic variations, particularly single nucleotide polymorphisms (SNPs), play a key role in influencing how individuals metabolize and respond to nutrients. These variations can impact overall health, disease risk, and responses to dietary interventions.
FTO Gene and Obesity	The FTO gene influences obesity and metabolic disorders. SNPs in the FTO gene affect appetite regulation, energy expenditure, and fat accumulation, increasing the risk of obesity. Personalized dietary strategies can help mitigate these risks.
MTHFR Gene and Folate Metabolism	The MTHFR gene is crucial for folate metabolism. SNPs in this gene can reduce the conversion of homocysteine to methionine, leading to higher cardiovascular disease risk. Individuals with MTHFR variants may benefit from increased folate intake or supplementation.
Lactase Persistence and Lactose Intolerance	Genetic variation determines lactase persistence, allowing some populations to digest lactose beyond infancy. In populations with low lactase persistence, lactose intolerance is common, requiring dietary adjustments to limit or avoid dairy products.
Nutritional Epigenetics	Nutritional epigenetics examines how nutrients and bioactive compounds affect gene expression through epigenetic mechanisms, such as DNA methylation and histone modifications, without altering the DNA sequence. These changes can have long-term health implications.
DNA Methylation	DNA methylation suppresses gene expression by adding a methyl group to cytosine in DNA. Nutrients like folate, vitamins B6, B12, and methionine support one-carbon metabolism, influencing methylation patterns and potentially reducing cardiovascular disease risk.
Histone Modifications	Nutrient-driven modifications to histones (e.g., acetylation, methylation) can promote or inhibit gene expression. For instance, polyphenols (e.g., resveratrol) from fruits can induce beneficial modifications that protect against inflammation and oxidative stress.
Methodologies in Nutritional Genomics	Nutritional genomics employs advanced methodologies, including genomics, transcriptomics, metabolomics, and bioinformatics, to unravel the complex interactions between diet, genetics, and health outcomes.
Genome-Wide Association Studies (GWAS)	GWAS identify genetic variants (SNPs) associated with nutrient metabolism, food intolerances, and disease risk. These studies are essential for developing personalized nutrition plans by linking SNPs to health traits and dietary responses.

Transcriptomics	Transcriptomics analyzes RNA to understand how gene expression changes in response to dietary components. Techniques like RNA sequencing (RNA-seq) offer insights into the modulation of gene expression by nutrients, such as omega-3 fatty acids, in health outcomes.
Metabolomics	Metabolomics studies small molecules (metabolites) in biological samples to capture the metabolic response to nutrients. This approach identifies biomarkers for disease risk and helps tailor dietary recommendations to optimize individual metabolic health.
Bioinformatics	Bioinformatics manages and interprets large-scale data from nutritional genomics, integrating GWAS, transcriptomics, and metabolomics to predict diet-gene interactions. Computational tools help develop personalized dietary interventions for disease prevention.

4. Personalized Nutrition and Diet Tailoring

Personalized nutrition, informed by nutritional genomics, offers a promising avenue for tailoring diets to an individual's genetic profile. This approach enables precise dietary recommendations designed to optimize health, improve athletic performance, and manage chronic diseases. The integration of genetic data into nutrition planning represents a paradigm shift from a one-size-fits-all diet approach to a more individualized strategy.

4.1 Improving Health Outcomes

One of the central goals of nutritional genomics is to enhance health outcomes by creating personalized dietary plans based on genetic predispositions. By understanding how specific genetic variants influence nutrient metabolism, healthcare professionals can offer more effective, targeted dietary recommendations that minimize disease risk and promote overall well-being. For example, individuals carrying polymorphisms in the APOE gene associated with hyperlipidemia and cardiovascular disease risk may benefit from a diet low in saturated fats and high in omega-3 fatty acids. Such tailored recommendations can help lower cholesterol levels, reduce inflammation, and ultimately decrease the risk of heart disease. Similarly, individuals with variations in the MTHFR gene may require increased intake of folate-rich foods to ensure proper metabolism of homocysteine and support cardiovascular health.

By integrating genetic insights, healthcare providers can customize nutritional interventions to prevent diseases before they manifest, leading to healthier outcomes over time. This approach fosters a proactive method to health management, where dietary interventions are preventive rather than reactive.

4.2 Enhanced Athletic Performance

Nutritional genomics also offers significant potential in enhancing athletic performance. By tailoring diets based on the genetic makeup of athletes, nutritionists can optimize nutrient intake to meet the specific metabolic demands of physical activity. Genetic testing enables the identification of variants related to energy metabolism, muscle function, recovery, and endurance, allowing for personalized adjustments in diet to improve performance and reduce injury risks. For example, athletes with variations in the ACTN3 gene, which influences fast-twitch muscle fibers, may benefit from higher protein intake to support muscle repair and recovery after intense exercise. Similarly, those with genetic tendencies toward lower fat oxidation rates may benefit from a diet with a specific macronutrient distribution—such as a higher carbohydrate intake for sustained energy during endurance events. Genetic profiling also helps identify nutrient requirements for better recovery. For instance, individuals with certain variations in IL6 or TNF-α genes may require anti-inflammatory diets rich in polyphenols and omega-3**s** to modulate inflammation and enhance recovery time. These personalized nutritional strategies contribute to peak performance while safeguarding long-term athletic health.

4.3 Chronic Disease Management

Personalized nutrition plays a pivotal role in chronic disease management, where individual genetic profiles can guide dietary interventions tailored to prevent or mitigate disease progression. Chronic conditions like diabetes, obesity, hypertension, and cardiovascular disease often have genetic components that interact with lifestyle and diet. By understanding these genetic predispositions, healthcare providers can design more effective dietary plans to manage symptoms and improve patients' quality of life. For instance, individuals with a genetic predisposition to insulin resistance, such as variations in the TCF7L2 gene, may be advised to consume a diet rich in whole grains, fiber, and low glycemic index (GI) carbohydrates. These dietary adjustments help to stabilize blood glucose levels and reduce the risk of developing type 2 diabetes. Additionally, individuals with genetic susceptibility to hypertension can benefit from a diet high in potassium and low in sodium, supported by personalized recommendations based on variants in the AGT gene.

Moreover, individuals at risk for obesity—due to genetic variations in genes like FTO—may require diets focused on controlled caloric intake, higher protein, and fiber to promote satiety and reduce excess weight gain. Tailored dietary approaches can also help manage inflammatory conditions, such as those influenced by genetic variations in IL6 or CRP genes, by recommending anti-inflammatory foods to lower systemic inflammation.

By leveraging genetic data, healthcare professionals can develop precise nutritional strategies that address the root causes of chronic diseases, improving long-term disease management and promoting healthier lifestyles. Personalized nutrition, in this context, offers the potential for precision medicine that adapts to an individual's unique genetic and nutritional needs.

5. Challenges and Considerations

While nutritional genomics offers promising avenues for improving health through personalized dietary strategies, the field faces several critical challenges. These challenges, ranging from ethical issues to the complexity of genetic-environmental interactions, need to be addressed to ensure the responsible and effective use of genetic information in nutrition.

5.1 Ethical Considerations

One of the primary concerns in nutritional genomics is the ethical implications of using genetic information in dietary recommendations. Issues related to privacy, informed consent, and potential discrimination must be carefully managed to protect individuals' rights.

- **Privacy**: As genetic data is highly personal, there is a risk of misuse if sensitive information is not safeguarded. Healthcare providers and genetic testing companies must implement stringent **data protection** measures to prevent unauthorized access or breaches of patient confidentiality.
- **Informed Consent**: Before undergoing genetic testing, individuals should be fully informed about the potential risks and benefits, including how their data will be used and stored. The consent process must be transparent to ensure that individuals are not unknowingly exposed to misuse or unintended consequences of their genetic information.
- **Discrimination**: There is concern that genetic data could be used to **discriminate** against individuals based on their predisposition to certain diseases. This could occur in various areas, including **employment** and **health insurance**. For instance, an individual with a genetic predisposition to obesity or heart disease might face discrimination in insurance premiums. To address this, **anti-discrimination legislation** must be put in place to protect individuals from the misuse of their genetic information.

The establishment of robust ethical guidelines is crucial for ensuring that genetic data is used responsibly in nutritional genomics, maintaining public trust and safeguarding individual rights.

5.2 Accessibility and Cost

A significant challenge in nutritional genomics is the issue of cost and accessibility. Genetic testing and subsequent personalized dietary interventions can be expensive, potentially limiting their availability to certain segments of the population. The high costs associated with genome sequencing, data analysis, and tailored nutritional counseling can restrict access, particularly for low-income and underserved communities.

- **Cost of Genetic Testing**: While the cost of genome sequencing has decreased over time, it remains prohibitive for many people. The price of a comprehensive genetic test ranges from hundreds to thousands of dollars, depending on the scope and depth of analysis required. Without affordable options, many individuals may be unable to access personalized nutrition plans that could benefit their health.
- **Equitable Access**: As the field of nutritional genomics advances, it is important to ensure that its benefits are **equitably distributed**. Public health efforts must focus on making genetic testing and personalized nutrition **more affordable and accessible** to all populations, particularly those at higher risk for chronic diseases. Additionally, healthcare systems need to integrate **personalized nutrition programs** into existing services to ensure broader reach.
- **Insurance Coverage**: Another barrier is the lack of **insurance coverage** for genetic testing and personalized nutrition interventions. Advocacy for policy changes that support **insurance reimbursement** for such services will be essential in promoting widespread adoption of nutritional genomics.

Efforts must be made to create affordable, scalable solutions that make personalized nutrition accessible to diverse populations, ensuring that everyone can benefit from advances in the field.

5.3 Complex Interactions

The relationship between genetics, diet, and health is highly complex, involving not only genetic variations but also environmental factors, lifestyle choices, and individual dietary habits. While nutritional genomics provides valuable insights into how genes influence nutrient metabolism, predicting individual responses to specific diets remains a challenge due to the intricate gene-environment interactions.

- **Gene-Diet Interactions**: Genetic variants alone do not determine how a person will respond to a particular diet. Environmental factors such as **exercise, stress**, and **exposure to toxins,** as well as lifestyle choices like

smoking or **alcohol consumption**, can significantly influence health outcomes. For example, an individual with a genetic predisposition to heart disease may reduce their risk through exercise and a healthy diet, but other environmental factors may negate these benefits.

- **Microbiome Influence**: Another layer of complexity is the role of the **gut microbiome** in modulating how the body processes nutrients. Emerging research suggests that the composition of an individual's microbiome can interact with both their genetics and diet, influencing nutrient absorption, metabolism, and immune response. Personalized nutrition must, therefore, consider not only genetic data but also microbiome health and other biological factors to create comprehensive dietary plans.
- **Polygenic Traits**: Many traits, such as **obesity, diabetes**, or **cardiovascular disease**, are influenced by multiple genes, making it difficult to predict an individual's response to a particular nutrient. **Polygenic risk scores** (PRS) are being developed to provide a more accurate prediction of disease risk by considering multiple genetic variants, but these scores are still in their infancy and may not yet offer a complete picture.
- **Nutritional Epidemiology**: Finally, while nutritional genomics provides a framework for understanding the genetic basis of nutrient metabolism, it should be integrated with **nutritional epidemiology**, which studies how various diets affect population health. This combination of genetic and population-level data is essential to formulating effective **public health guidelines** that accommodate both personalized and population-based approaches.

As research progresses, understanding these complex interactions will be key to improving the precision and effectiveness of dietary recommendations, ensuring they are not only genetically informed but also contextually appropriate for each individual's environment and lifestyle.

6. Future Directions

The future of nutritional genomics promises to transform the way we understand and approach dietary recommendations. One of the key future directions is the integration of multi-omic approaches, combining genomics, transcriptomics, proteomics, and metabolomics to create a comprehensive model of human nutritional responses. By analyzing data across multiple biological layers, researchers will gain a more nuanced understanding of how genes, proteins, and metabolites interact with dietary components to influence health outcomes.

6.1 Integrating Multi-Omic Approaches

- **Genomics**: Provides insights into how genetic variants affect nutrient metabolism and disease risk.
- **Transcriptomics**: Investigates gene expression changes in response to specific diets or nutrients, highlighting the regulatory effects of diet on health.
- **Proteomics**: Studies the proteins and enzymes involved in nutrient absorption, digestion, and metabolism, shedding light on how individual proteins respond to dietary interventions.
- **Metabolomics**: Measures the metabolites resulting from metabolic processes, offering real-time insights into how an individual's biochemistry is affected by their diet and genetic profile.

Integrating these multi-omic datasets will provide a holistic view of nutrition and health, revealing the complex interactions that drive dietary responses and identifying specific biomarkers of metabolic health. This knowledge will enable the development of precision nutrition plans that not only consider genetic factors but also how they are modulated by diet, lifestyle, and environmental factors.

6.2 Advancements in Artificial Intelligence and Machine Learning

As nutritional genomics generates increasingly large and complex datasets, artificial intelligence (AI) and machine learning will play a pivotal role in making sense of the data. These technologies are well-suited to analyze intricate patterns and interactions that might be difficult for humans to detect. For instance, AI can be used to predict how genetic variations will influence responses to specific nutrients and design personalized dietary plans accordingly.

- **Predictive Models**: Machine learning algorithms can analyze multi-omic data and predict individual responses to diet. These models will provide more accurate and actionable dietary recommendations tailored to each person's genetic makeup, helping to prevent disease and promote long-term health.
- **Personalized Nutrition Platforms**: In the near future, AI-powered platforms could integrate genetic data, lifestyle factors, and health metrics to provide real-time personalized nutrition advice. These platforms would allow users to adjust their diets dynamically, based on ongoing genetic and health data monitoring.

The combination of AI-driven analytics with nutritional genomics will enhance the precision and scalability of personalized nutrition, making it accessible to larger populations while maintaining individualized specificity.

6.3 Collaboration and Policy Development

The successful translation of nutritional genomics into practical applications requires collaboration across multiple disciplines, including researchers, healthcare providers, policymakers, and industry stakeholders.

- **Research**: Ongoing studies are needed to validate the clinical utility of nutritional genomics in diverse populations. There is also a need for more long-term studies to assess how personalized nutrition affects disease prevention and overall health outcomes over time.
- **Healthcare**: Healthcare professionals will need training in nutritional genomics to effectively integrate genetic data into dietary recommendations. Additionally, genetic counseling may become a routine part of healthcare, helping patients understand their genetic profiles and how they relate to nutrition.
- **Policy**: Policymakers must work to ensure that advancements in nutritional genomics are ethically regulated and accessible to all, regardless of socioeconomic background. This includes addressing issues of data privacy, equity in access to genetic testing, and the development of ethical frameworks to prevent misuse of genetic information.

6.4 Expansion of Personalized Nutrition

As nutritional genomics continues to evolve, there will likely be a shift toward personalized nutrition as the standard approach to dietary recommendations.

- **Consumer Products**: Personalized nutrition services, such as genetic testing kits paired with dietary advice, are becoming increasingly available to consumers. In the future, it is expected that more consumer products will be developed to cater to individual genetic profiles, from supplements to meal planning services.
- **Public Health**: On a broader scale, public health initiatives may begin incorporating genetic data to create more tailored dietary guidelines for different population groups. Personalized nutrition could eventually help combat chronic diseases such as diabetes, cardiovascular disease, and obesity, which have strong genetic and dietary components.

7. Conclusion

Nutritional genomics stands at the forefront of a revolution in dietary science, offering unprecedented opportunities to tailor diets to individual genetic profiles.

By understanding the intricate relationships between genetics, nutrition, and health, we can develop dietary strategies that are not only preventive but also therapeutic. Personalizing diets in this way has the potential to enhance overall health outcomes, prevent chronic diseases, and promote optimal well-being throughout life. Despite the exciting promise of this field, several challenges remain. Addressing ethical concerns, ensuring equitable access to genetic testing and personalized interventions, and navigating the complexities of gene-diet interactions are all critical steps toward realizing the full potential of nutritional genomics. By fostering continued research, collaboration between stakeholders and the development of ethical and practical frameworks, the field can evolve to deliver precision nutrition that is both effective and accessible.

As the science of nutritional genomics advances, it will play an essential role in shaping the future of nutrition and healthcare, allowing us to move beyond one-size-fits-all recommendations toward personalized nutrition that accounts for individual genetic and biological diversity. Through these advances, we can harness the power of genetics to create more effective, sustainable, and inclusive dietary strategies that improve human health and well-being at both the individual and population levels.

References

Afman, L., & Müller, M. (2006). Nutrigenomics: From molecular nutrition to prevention of disease. Journal of the American Dietetic Association, 106(4), 569-576.

Ahmad, T., et al. (2015). Nutrigenomics: A path to future personalized diet. Journal of Nutrition and Metabolism, 2015, 1-6.

Basu, S., et al. (2013). Role of epigenetics in health and disease: Nutrigenomics and cancer. Food and Chemical Toxicology, 59, 164-171.

Clarke, G., Sandhu, K. V., Paul, A. M., Varela, C., et al. (2016). Gut microbiota: The neglected endocrine organ. Molecular Endocrinology, 30(2), 247–267.

Corella, D., & Ordovás, J. M. (2017). Nutrigenomics in cardiovascular medicine. Circulation Research, 122(8), 1-13.

Fenech, M. (2012). Nutrigenomics and Nutrigenetics: The next wave in nutrition research. Environmental Health Perspectives, 120(1), 32-38.

Ferguson, L. R., & De Caterina, R. (2019). Nutrition and gene expression: Introduction. Genes & Nutrition, 14(1), 1-2.

Ferguson, L. R., et al. (2016). Nutrigenomics, epigenetics and gut microbiota: Translating human nutrition into personalized health. Journal of Translational Medicine, 14(1), 1-10.

Horne, J. R., et al. (2019). Dietary influences on epigenetic regulators in cancer. Advances in Nutrition, 10(2), 247-260.

Kaikkonen, M. U., et al. (2014). Epigenetic mechanisms in atherosclerosis and diabetes. Nature Reviews Endocrinology, 10(11), 688-698.

Kaput, J., & Rodriguez, R. L. (2004). Nutritional genomics: the next frontier in the postgenomic era. Physiological Genomics, 16(2), 166-177.

Kim, Y. I. (2005). Nutritional epigenetics: Impact of folate deficiency on DNA methylation and colon cancer risk. The Journal of Nutrition, 135(11), 2703-2709.

Kohlmeier, M. (2015). Nutrigenetics: Applying the science of personal nutrition. Elsevier.

Lidder, S., & Webb, A. J. (2013). Vascular effects of dietary nitrate (as found in green leafy vegetables and beetroot) via the nitrate-nitrite-nitric oxide pathway. British Journal of Clinical Pharmacology, 75(3), 677-696.

Liu, Y., et al. (2018). Nutrigenomics and nutrigenetics: Advancing the science of individualized nutrition. Journal of Nutritional Biochemistry, 52, 1-8.

Livingstone, K. M., et al. (2016). Nutrigenomics and personalized nutrition: Insights into human metabolism. Genes & Nutrition, 11(1), 1-9.

Manore, M. M., et al. (2017). Nutritional genomics, the gut microbiome, and precision nutrition: Moving toward personalized nutrition. Journal of the Academy of Nutrition and Dietetics, 117(12), 1945-1955.

Maruvada, P., et al. (2020). Perspective: Interpreting Diet-Gene Interactions in Personalized Nutrition and Dietetics. Advances in Nutrition, 11(1), 12-25.

Mathers, J. C. (2017). Nutrigenomics in the postgenomic era: Understanding the role of nutrition in shaping health and disease. Nutrition Bulletin, 42(3), 245-253.

Milenkovic, D., et al. (2017). Nutrigenomics: New perspectives for personalized nutrition and precision health. Journal of Clinical Medicine, 6(8), 64-78.

Nielsen, D. E., & El-Sohemy, A. (2014). Nutrigenetics and personalized nutrition: Science and concept. Genes & Nutrition, 9(5), 1-8.

Ordovás, J. M., & Ferguson, L. R. (2017). Nutrigenomics and personalized nutrition: science and concept. Genes & Nutrition, 12(1), 1-2.

Ordovás, J. M., & Mooser, V. (2004). Nutrigenomics and nutrigenetics. Current Opinion in Lipidology, 15(2), 101-108.

Ordovás, J. M., et al. (2018). Personalised nutrition and health. BMJ, 361(1), 1-10.

Rezvani, Z., & Backes, A. C. (2018). The relationship between nutrigenomics, epigenetics, and personalized nutrition. Journal of Nutritional Science, 5(12), 60-72.

Rimbach, G., et al. (2010). Nutrigenetics and nutrient requirements. Journal of Nutrition, 140(12), 360S-366S.

Roberts, R., et al. (2020). Nutrigenomics: An overview of the science and applications in clinical nutrition. Nutrients, 12(3), 778-790.

Tollefsbol, T. O. (Ed.). (2011). Epigenetics in Human Disease. Academic Press.

Zeevi, D., Korem, T., Zmora, N., & Segal, E. (2015). Personalized nutrition by prediction of glycemic responses. Cell, 163(5), 1079-1094.

Zeng, C., et al. (2019). Emerging role of epigenetics in cardiovascular diseases. Journal of Medical Genetics, 56(1), 21-28.

6

Probiotics and Prebiotics: Modulating Fish Gut Health Through Nutrition

1. Introduction

Aquaculture is a rapidly expanding industry, playing a crucial role in addressing the global demand for high-quality protein sources as wild fish stocks face depletion. With aquaculture production on the rise, ensuring the health and productivity of farmed fish has become increasingly important. The health of the fish gut is central to achieving these objectives, as the gut is responsible not only for nutrient digestion and absorption but also for maintaining a strong immune defense. A well-functioning gut leads to better growth performance, reduced disease susceptibility, and improved overall fish welfare.

However, maintaining optimal gut health in aquaculture systems presents challenges. The intensive nature of fish farming can lead to stressors that compromise the gut environment, including changes in water quality, fluctuating diet compositions, and exposure to pathogens. Gut health issues often manifest in poor feed conversion ratios, slower growth, and higher mortality rates, all of which negatively impact aquaculture productivity and sustainability. Therefore, there is a growing focus on finding innovative and natural solutions to improve the gut health of farmed fish.

In recent years, the use of probiotics and prebiotics in fish nutrition has gained significant attention as an effective and eco-friendly strategy to modulate gut microbiota and enhance fish health. These functional feed additives are considered a natural alternative to antibiotics and other pharmaceuticals, which have historically been used to manage diseases and promote growth in aquaculture but come with risks of antibiotic resistance and environmental contamination.

- **Probiotics** are live beneficial microorganisms that, when administered through feed or water, can colonize the fish's gastrointestinal tract, promoting a healthy balance of gut bacteria. They contribute to the host's health by inhibiting pathogenic bacteria, improving digestion, and modulating the immune system.

- **Prebiotics** are indigestible dietary components that selectively stimulate the growth and activity of beneficial gut microbes. Unlike probiotics, which introduce new microorganisms, prebiotics act as a substrate to enhance the growth of the existing beneficial microbial populations within the fish gut.

These functional additives are being recognized for their ability to enhance gut health, improve digestion, and strengthen immune responses. By doing so, they contribute to disease resistance, reduce the need for antibiotics, and promote overall fish growth and performance. Furthermore, probiotics and prebiotics support sustainable aquaculture practices by reducing environmental impacts and aligning with global efforts to ensure responsible fish farming methods. The mechanisms by which probiotics and prebiotics exert their positive effects on fish gut health are varied and multifaceted. Probiotics can outcompete harmful bacteria, produce antimicrobial substances, and stimulate immune responses, while prebiotics promote the proliferation of beneficial gut microbes and enhance nutrient absorption. The combined use of both—known as **synbiotics**—has shown synergistic effects, making this approach even more promising for the aquaculture industry.

In this chapter, we will delve into the scientific understanding of probiotics and prebiotics, exploring how they modulate fish gut health, their mechanisms of action, and their potential to improve growth performance and disease resistance. We will also examine the current research surrounding these functional feed additives and their role in promoting sustainable aquaculture practices. As the industry continues to evolve, the integration of probiotics and prebiotics into fish nutrition strategies is likely to become a cornerstone of future aquaculture development, offering a natural, effective, and sustainable solution to the challenges of maintaining fish health in farming systems.

2. Understanding Probiotics and Prebiotics

In aquaculture, the health and performance of farmed fish are closely linked to the composition and activity of the gut microbiota, which plays a critical role in digestion, nutrient absorption, and immunity. Gut health is influenced by various environmental and dietary factors, and disruptions in the balance of gut microbiota can lead to health issues, including reduced growth rates and increased susceptibility to diseases. To promote gut health and improve overall fish well-being, probiotics and prebiotics are increasingly used as functional feed additives in aquaculture systems.

2.1 Probiotics: Beneficial Microorganisms

Probiotics are live microorganisms that, when administered in adequate amounts, confer health benefits to the host, particularly by improving the balance of the gut microbiota. The term "probiotic" is derived from the Greek words "pro," meaning "for," and "biotic," meaning "life." These beneficial microbes help maintain a healthy gut environment by modulating the microbial population in favor of beneficial bacteria, thereby enhancing the fish's overall health and performance. In aquaculture, probiotics are typically introduced into the fish's diet via feed or water, and they include various strains of bacteria, yeast, or fungi. The most commonly used probiotic species in fish nutrition are from the genera Lactobacillus**,** Bifidobacterium**,** Bacillus, Enterococcus, and Saccharomyces. Each of these strains has unique properties that contribute to gut health and overall fish well-being. The administration of probiotics in aquaculture has been associated with numerous health benefits, including improved digestive efficiency, enhanced immune function, and disease resistance. Probiotics work through various mechanisms that positively affect the gut microbiota and the host's immune system, resulting in better growth performance and reduced mortality. These mechanisms are explored in detail below:

Mechanisms of Action

Competitive Exclusion

One of the key mechanisms by which probiotics exert their beneficial effects is through competitive exclusion. Probiotics compete with harmful pathogens for both nutrients and adhesion sites on the intestinal mucosa, thereby reducing the chances of pathogen colonization. By occupying these adhesion sites, probiotics prevent harmful bacteria from attaching to the intestinal lining, limiting their ability to establish themselves and cause infections. Competitive exclusion is especially important in aquaculture systems, where fish are constantly exposed to potential pathogens present in the water and feed. Probiotics help maintain a healthy microbial balance by ensuring that beneficial bacteria outnumber harmful ones, reducing the risk of disease outbreaks.

Antimicrobial Production

Certain probiotic strains are capable of producing antimicrobial substances, such as bacteriocins, hydrogen peroxide, and organic acids, which inhibit the growth of harmful bacteria. These antimicrobial compounds can target and neutralize pathogenic bacteria, preventing them from proliferating in the fish gut.

- **Bacteriocins** are proteinaceous toxins produced by bacteria that inhibit the growth of closely related bacterial species, providing a natural defense mechanism against pathogenic microbes.

- **Hydrogen peroxide** disrupts the cellular processes of harmful bacteria, reducing their viability in the gut environment.
- **Organic acids** (e.g., lactic acid and acetic acid) lower the pH of the gut, creating an inhospitable environment for pathogens that thrive in neutral or alkaline conditions. This reduction in gut pH also promotes the growth of acid-tolerant beneficial bacteria, further supporting gut health.

Modulation of Immune Responses

Probiotics play a significant role in modulating the host's immune system. They can enhance the activity of immune cells, such as macrophages and neutrophils, and increase the expression of immune-related genes. Probiotics stimulate the production of cytokines and other signaling molecules that activate immune responses, thereby improving the fish's ability to fight off infections. By promoting immune cell activation and enhancing the production of antimicrobial peptides and lysozyme, probiotics bolster the fish's innate immune defenses. This heightened immune response is especially beneficial in preventing infections from common aquaculture pathogens, such as Vibrio, Aeromonas, and Edwardsiella species. Moreover, probiotics can influence the adaptive immune system by enhancing the production of antibodies, such as IgM, which play a crucial role in neutralizing pathogens. This immune-modulating effect not only improves disease resistance but also reduces the severity and duration of infections when they occur.

Improved Digestive Function

Probiotics contribute to improved digestion by secreting digestive enzymes that aid in the breakdown of complex nutrients in the gut. For example, certain probiotic strains produce enzymes such as amylase, protease, and lipase, which enhance the digestion of carbohydrates, proteins, and fats, respectively. By improving the digestion and absorption of nutrients, probiotics can enhance feed conversion efficiency, leading to better growth performance in fish. In addition, probiotics can stimulate the production of short-chain fatty acids (SCFAs), which serve as an energy source for gut epithelial cells. These SCFAs, such as acetate, propionate, and butyrate, are produced during the fermentation of dietary fibers by beneficial bacteria. They not only provide energy to the gut lining but also play a role in maintaining gut health by reducing inflammation and supporting the growth of beneficial microbes.

Maintenance of Gut Integrity

The integrity of the intestinal barrier is critical for preventing the entry of pathogens and toxins into the bloodstream. Probiotics help maintain and strengthen the gut epithelial barrier by promoting the production of mucus and enhancing tight junctions between intestinal cells. Tight junctions are protein complexes that seal the spaces between epithelial cells, preventing harmful substances from passing through the gut lining. By improving the structural integrity of the gut, probiotics reduce gut permeability, also known as "leaky gut," which is a condition where toxins and pathogens can cross the gut barrier and enter the circulatory system. Maintaining a strong gut barrier not only protects fish from infections but also minimizes inflammatory responses, contributing to overall health and well-being.

2.2 Prebiotics: Feeding the Beneficial Microbiota

Prebiotics are non-digestible food components that selectively promote the growth and activity of beneficial microorganisms in the gut. Unlike probiotics, which introduce beneficial microbes into the host, prebiotics act as a food source for the existing beneficial gut bacteria, encouraging their proliferation and activity. These substances, typically composed of complex carbohydrates, support the health of the gut microbiota and, in turn, enhance fish health and productivity. In aquaculture, prebiotics are commonly used to optimize gut health, improve nutrient utilization, and enhance immune function, all of which contribute to better growth and disease resistance in farmed fish.

Common prebiotics used in fish nutrition include:

- **Oligosaccharides** (e.g., fructooligosaccharides, mannan oligosaccharides)
- **Inulin**
- **Galactooligosaccharides**
- **Xylooligosaccharides**

These compounds are often added to fish feed formulations to promote gut health and provide sustainable alternatives to antibiotics and synthetic growth promoters.

Mechanisms of Action

Selective Fermentation

One of the most critical mechanisms by which prebiotics promote gut health is through selective fermentation. Prebiotics serve as a substrate for specific beneficial bacteria, such as Lactobacillus and Bifidobacterium, which thrive on these complex carbohydrates. Upon fermentation, these bacteria produce short-chain fatty acids (SCFAs), including acetate, propionate, and butyrate, which have a range of beneficial effects on gut health and fish performance.

SCFAs play several roles in maintaining gut health

- **Antimicrobial Properties**: SCFAs lower the gut pH, creating an unfavorable environment for pathogenic bacteria, which prefer more neutral or alkaline conditions. By reducing the gut pH, SCFAs limit the proliferation of harmful bacteria and promote the growth of beneficial microbes.
- **Energy Source for Gut Epithelial Cells**: SCFAs, particularly butyrate, provide an essential energy source for the cells lining the gut, known as **enterocytes**. This energy supports cellular functions, promotes cell regeneration, and contributes to the maintenance of a healthy gut barrier.
- **Inhibition of Pathogens**: In addition to their indirect antimicrobial effects via pH reduction, SCFAs can also inhibit pathogens directly by interfering with their metabolic pathways and reducing their ability to colonize the gut.

The production of SCFAs through prebiotic fermentation plays a crucial role in creating a balanced and healthy gut microbiome, which in turn enhances overall fish health.

Enhanced Gut Barrier Function

Another vital effect of prebiotics is the improvement of gut barrier function. The gut lining acts as a barrier that prevents pathogens and toxins from entering the fish's bloodstream while allowing the absorption of essential nutrients. Prebiotics support the integrity of this barrier by promoting the production of SCFAs, which help maintain the tight junctions between gut epithelial cells. Tight junctions are protein structures that bind intestinal cells together and prevent harmful substances from leaking into the circulatory system, a condition commonly referred to as leaky gut. By strengthening these tight junctions, prebiotics reduce gut permeability, lower inflammation, and prevent the entry of pathogens and toxins into the body.

Additionally, prebiotics stimulate the production of mucus in the gut, which acts as an additional protective layer that traps harmful microbes and toxins, preventing them from reaching the gut lining. This dual protection — enhanced mucus production and stronger tight junctions — helps maintain a healthy and functional gut barrier, which is essential for preventing infections and promoting nutrient absorption.

Improved Immune Modulation

Prebiotics also contribute to enhanced immune modulation by interacting with the gut-associated lymphoid tissue (GALT), a critical component of the immune system located in the intestines. Prebiotics influence the immune system in several ways:

- **Increased Antibody Production**: Prebiotics have been shown to enhance the production of **immunoglobulins** (antibodies), particularly **IgM**, which plays a vital role in the immune response by neutralizing pathogens and preventing infections.
- **Activation of Immune Cells**: Prebiotics can stimulate the activity of immune cells such as **macrophages** and **lymphocytes**, which are responsible for identifying and eliminating pathogens. These immune cells are critical in the defense against bacterial and viral infections, and their activation contributes to overall disease resistance in fish.
- **Reduced Inflammation**: By promoting a balanced gut microbiota and enhancing the production of SCFAs, prebiotics help reduce inflammation in the gut. Inflammation is often associated with poor gut health and can lead to a weakened immune response. By reducing inflammation, prebiotics contribute to a healthier and more effective immune system.

Through these immune-modulating effects, prebiotics support the fish's ability to resist infections, reducing the need for antibiotics and other chemical treatments in aquaculture systems.

Improved Nutrient Utilization

Prebiotics indirectly improve nutrient digestibility and absorption by promoting the growth of beneficial bacteria that aid in digestion. Beneficial gut microbes produce enzymes that help break down complex dietary components, making nutrients more available for absorption. This process contributes to better feed conversion efficiency and enhances overall growth performance in fish. For instance, mannan oligosaccharides (MOS) can promote the growth of beneficial bacteria that assist in the digestion of proteins and carbohydrates, improving the bioavailability of essential nutrients such as amino acids, vitamins, and minerals. By enhancing nutrient absorption, prebiotics not only support healthy growth but also help optimize feed utilization, reducing feed waste and promoting more sustainable aquaculture practices.

3. Probiotics and Prebiotics in Aquaculture: A Synergistic Approach

In modern aquaculture, there is a growing interest in using functional feed additives that not only enhance fish growth but also improve gut health and disease resistance in a sustainable manner. One of the most promising approaches is the use of synbiotics, which combine probiotics and prebiotics in aquaculture diets. This synergistic strategy leverages the benefits of both components to enhance gut health, nutrient utilization, immune function, and overall fish performance. Probiotics and prebiotics work through different but complementary mechanisms. When used together, they create a more favorable environment for beneficial microbes, strengthen the immune system, and support optimal growth and health in fish. This approach offers a more comprehensive solution for aquaculture, addressing both the prevention of disease and the optimization of growth, while reducing the reliance on antibiotics and chemical treatments.

3.1 Increased Probiotic Colonization

One of the primary advantages of combining probiotics and prebiotics is the enhancement of probiotic colonization in the fish gut. While probiotics provide beneficial microorganisms, their colonization and persistence in the gut can sometimes be limited due to the competitive environment and the need for a suitable food source. Prebiotics help overcome this challenge by serving as a substrate for these probiotics, ensuring their survival and promoting their proliferation.

- **Food Source for Probiotics**: Prebiotics, typically non-digestible fibers like oligosaccharides, inulin, and mannan oligosaccharides (MOS), provide a selective food source for probiotics, allowing them to thrive in the gut. The fermentation of prebiotics by probiotic bacteria such as **Lactobacillus** and **Bifidobacterium** leads to the production of **short-chain fatty acids (SCFAs)**, which further support gut health and reduce the colonization of harmful pathogens.
- **Improved Gut Microbiota Balance**: The combined action of probiotics and prebiotics enhances the balance of the gut microbiota, favoring the growth of beneficial bacteria over harmful pathogens. This leads to a healthier gut environment, which supports efficient digestion, nutrient absorption, and immune function in fish.
- **Longevity of Probiotic Strains**: Prebiotics help probiotics survive through challenging conditions, such as changes in diet or environmental stressors. By promoting the growth and colonization of beneficial bacteria, prebiotics ensure that probiotics continue to exert their health-promoting effects over an extended period.

3.2 Enhanced Immune Response

Synbiotics play a crucial role in improving the immune response of fish, leading to better disease resistance and overall health. Both probiotics and prebiotics individually have immunomodulatory properties, but their combination amplifies these effects, providing a stronger and more sustained immune response.

- **Stimulation of Immune Cells**: Probiotics have been shown to activate various components of the immune system, including **macrophages, neutrophils**, and **lymphocytes**, which are involved in pathogen recognition and elimination. Prebiotics further enhance this process by promoting the production of **protective antibodies** (such as **IgM**) and supporting the activity of immune cells.
- **Synergistic Immune Modulation**: The combined use of probiotics and prebiotics enhances the production of **cytokines** and **anti-inflammatory agents**, reducing the likelihood of harmful inflammatory responses that can compromise gut health and overall fish well-being. This synergy also improves the fish's ability to respond to pathogen challenges, ensuring a more robust defense against common aquaculture diseases such as bacterial infections, parasitic infestations, and viral outbreaks.
- **Reduced Disease Incidence**: With a stronger immune system, fish treated with synbiotics are more resistant to infections, leading to lower disease incidence and reduced reliance on antibiotics or chemical treatments. This promotes a healthier and more sustainable aquaculture environment, contributing to long-term productivity.

3.3 Improved Growth Performance

Another significant benefit of synbiotics in aquaculture is their ability to improve growth performance. By enhancing nutrient digestion and absorption, synbiotics contribute to more efficient feed utilization, leading to better growth rates, improved feed conversion ratios (FCR), and overall productivity.

- **Enhanced Nutrient Absorption**: Probiotics contribute to improved digestive function by secreting enzymes that break down complex nutrients, while prebiotics promote the growth of beneficial bacteria that aid in nutrient fermentation. This combined effect leads to better digestion and absorption of essential nutrients, including **proteins, carbohydrates**, and **lipids**, resulting in higher growth rates.
- **Improved Feed Conversion Ratios (FCR)**: The combination of probiotics and prebiotics enhances the efficiency of nutrient utilization, which directly impacts the **feed conversion ratio**. A lower FCR means that fish are able to convert feed into body mass more efficiently, which reduces feed waste and increases the sustainability of aquaculture operations.
- **Reduced Feed Costs**: By improving feed efficiency and reducing feed waste, synbiotics can help lower the overall cost of production. This is particularly important in commercial aquaculture, where feed costs

constitute a significant portion of operational expenses. Synbiotics can contribute to more cost-effective and environmentally sustainable production systems.

3.4 Environmental and Economic Benefits

The use of synbiotics in aquaculture not only enhances fish health and growth but also provides broader environmental and economic benefits. By reducing the need for antibiotics and chemical treatments, synbiotics contribute to more sustainable aquaculture practices, which are essential for long-term industry growth.

- **Reduced Antibiotic Use**: The enhanced disease resistance provided by synbiotics reduces the reliance on antibiotics, which is crucial for preventing the development of **antibiotic-resistant bacteria**. This not only benefits fish health but also addresses global concerns about antimicrobial resistance in aquaculture.
- **Environmental Sustainability**: By improving nutrient utilization and reducing feed waste, synbiotics help minimize the environmental impact of aquaculture operations. Lower feed waste means reduced nutrient pollution in water bodies, which is critical for maintaining water quality and protecting aquatic ecosystems.
- **Cost-Effective Production**: With improved growth performance, feed efficiency, and disease resistance, synbiotics offer a cost-effective solution for aquaculture producers. The long-term benefits of reduced disease outbreaks, lower feed costs, and minimal environmental impact make synbiotics a valuable tool for the economic sustainability of the industry.

4. The Impact of Probiotics and Prebiotics on Fish Gut Health

Fish gut health is a critical factor influencing overall growth performance, disease resistance, and nutrient absorption. The gastrointestinal microbiota, which consists of diverse microbial species, plays a central role in these processes. Probiotics and prebiotics have emerged as powerful tools in aquaculture for modulating the gut microbiota, promoting beneficial bacteria, and suppressing harmful pathogens. This modulation contributes to improved digestive efficiency, disease resistance, and better nutrient absorption.

4.1 Modulation of Gut Microbiota

The gut microbiota refers to the community of microorganisms inhabiting the digestive tract of fish, which significantly impacts nutrient digestion, immune function, and pathogen defense. Probiotics and prebiotics work synergistically to influence the composition and functionality of this microbiota. They achieve this by encouraging the growth of beneficial bacteria while simultaneously creating conditions that limit the proliferation of harmful pathogens. The modulation of the gut microbiota by these functional additives offers several key benefits:

1. **Increase in Beneficial Bacteria**: Probiotics introduce and sustain populations of live beneficial microorganisms such as Lactobacillus, Bifidobacterium, Enterococcus, and Bacillus in the gut. These bacterial genera play essential roles in promoting digestion, competing with pathogens for adhesion sites, and stimulating immune responses. Prebiotics, on the other hand, act as a selective food source for these beneficial bacteria, promoting their growth and activity. Common prebiotics like fructooligosaccharides (FOS) and mannan oligosaccharides (MOS) are fermented by beneficial bacteria, resulting in the production of metabolites such as short-chain fatty acids (SCFAs), which create a favorable environment for healthy microbial communities.
2. **Suppression of Pathogens**: One of the primary benefits of a probiotic and prebiotic-enriched diet is the suppression of harmful pathogens such as Vibrio spp., Aeromonas spp., and Edwardsiella spp., which are known to cause significant losses in aquaculture due to disease outbreaks. Probiotics prevent pathogen colonization through competitive exclusion, whereby beneficial bacteria outcompete pathogens for available nutrients and attachment sites on the gut lining. Additionally, certain probiotics produce antimicrobial substances such as bacteriocins, organic acids, and hydrogen peroxide, which directly inhibit pathogen growth. Prebiotics also help by creating a more acidic gut environment through the production of SCFAs, which lower gut pH and create unfavorable conditions for pathogenic bacteria.
3. **Enhanced Microbial Diversity and Stability**: A diverse gut microbiota is more resistant to disruptions, such as dietary changes, environmental stress, and disease challenges. Fish fed diets supplemented with probiotics and prebiotics exhibit a more diverse and stable microbial community. Increased microbial diversity offers enhanced metabolic capabilities, improving the breakdown of complex nutrients and enhancing nutrient absorption. Stability in the gut microbiota is critical for maintaining homeostasis, reducing the likelihood of dysbiosis (an imbalance in microbial populations), and protecting the fish from opportunistic pathogens that may exploit any disturbance in the gut environment.

4. **Improved Digestive Efficiency**: The modulation of the gut microbiota with probiotics and prebiotics leads to better digestive efficiency. Beneficial bacteria contribute to the breakdown of complex dietary components, such as fibers and starches, which would otherwise be indigestible. The fermentation of these components by gut bacteria results in the production of SCFAs that not only provide energy to gut epithelial cells but also improve nutrient absorption. Probiotic bacteria also produce digestive enzymes that enhance the breakdown and assimilation of nutrients, ensuring that fish derive the maximum benefit from their feed. This translates into better feed conversion ratios (FCR), higher weight gain, and improved overall growth performance.

Studies on Gut Microbiota Modulation

Several studies have demonstrated the positive effects of probiotics and prebiotics on fish gut microbiota. For example:

- A study on rainbow trout (*Oncorhynchus mykiss*) showed that supplementation with probiotics increased the relative abundance of Lactobacillus and Bifidobacterium, resulting in improved gut health and resistance to infections.
- In Nile tilapia (*Oreochromis niloticus*), dietary supplementation with prebiotics, such as inulin and FOS, significantly enhanced the populations of beneficial gut bacteria, leading to improved growth performance and disease resistance against bacterial pathogens such as *Aeromonas hydrophila*.
- In common carp (*Cyprinus carpio*), feeding probiotics such as *Bacillus subtilis* improved digestive efficiency and enhanced the immune response, reducing the risk of infection by harmful bacteria like *Edwardsiella tarda*.

4.2 Enhancement of Immune Function

The immune system plays a crucial role in defending fish against pathogens and maintaining overall health. The gastrointestinal (GI) tract, which is a primary site of interaction with the external environment, serves as the first line of defense against ingested pathogens. The gut-associated lymphoid tissue (GALT) is an important component of the immune system in fish, and its function is closely linked to the health of the gut microbiota. Probiotics and prebiotics have been shown to enhance the immune responses of fish by influencing both the innate and adaptive immune systems, thereby improving resistance to diseases and increasing survival rates in aquaculture.

Modulation of Innate and Adaptive Immunity

Innate immunity is the first and most immediate line of defense against infections and is particularly important in fish, as their adaptive immune responses are slower to develop than in mammals. The innate immune system includes physical barriers (such as the gut epithelium), immune cells (such as macrophages and neutrophils), and antimicrobial substances like lysozymes. Probiotics and prebiotics boost these innate defenses through several mechanisms:

1. **Stimulation of Lysozyme Production**: Lysozyme is an enzyme that breaks down the cell walls of gram-positive bacteria, serving as a key antimicrobial agent in fish. Probiotic supplementation has been shown to significantly increase lysozyme activity in fish, enhancing their ability to neutralize bacterial infections. For example, *Lactobacillus rhamnosus* and *Bacillus subtilis*, two common probiotic strains used in aquaculture, have been demonstrated to boost lysozyme production in species like rainbow trout (*Oncorhynchus mykiss*) and common carp (*Cyprinus carpio*). Enhanced lysozyme levels contribute to faster pathogen clearance and reduced disease severity, which is critical for improving fish survival in the high-density environments of aquaculture systems.
2. **Activation of Phagocytic Cells**: Macrophages and neutrophils are key phagocytic cells in the innate immune system that engulf and destroy pathogens. Probiotics can enhance the activity of these cells, leading to more effective pathogen elimination. Studies have demonstrated that fish fed with probiotic-supplemented diets exhibit higher levels of phagocytic activity, meaning that their immune cells are more efficient at identifying and removing harmful bacteria and viruses. The activation of phagocytic cells by probiotics like Bacillus spp. and *Saccharomyces cerevisiae* has been shown to improve immune responses in species such as tilapia (*Oreochromis niloticus*) and Asian seabass (*Lates calcarifer*), offering better protection against pathogens like *Aeromonas hydrophila* and Vibrio spp..
3. **Production of Antimicrobial Substances**: Some probiotics can directly produce antimicrobial substances, such as bacteriocins and organic acids, that inhibit the growth of pathogenic bacteria. For example, Lactobacillus spp. and Enterococcus spp. are known to secrete organic acids that lower the gut pH, creating an environment unfavorable to pathogens like Vibrio and Edwardsiella.

On the other hand, adaptive immunity provides a more specific and long-lasting response to pathogens. Probiotics and prebiotics enhance adaptive immunity by influencing the production of antibodies and the activation of immune cells that have a memory of specific pathogens:

1. **Enhanced Antibody Production**: Antibodies play a critical role in the adaptive immune response, as they bind to specific antigens on pathogens, marking them for destruction by immune cells. Prebiotics like fructooligosaccharides (FOS) and mannan oligosaccharides (MOS) have been shown to increase the production of antibodies, such as immunoglobulin M (IgM), in fish. Research has demonstrated that fish fed diets containing both probiotics and prebiotics (i.e., synbiotics) exhibit significantly higher levels of IgM in their bloodstreams, which enhances their ability to neutralize infections. For example, studies on Nile tilapia and rainbow trout have shown that synbiotics increase serum IgM levels, resulting in improved resistance to bacterial pathogens like Aeromonas spp..
2. **Upregulation of Immune-Related Genes**: Probiotics and prebiotics can influence gene expression related to the immune system. For example, studies have shown that fish fed diets supplemented with probiotics experience upregulation of immune-related genes, including those involved in the production of cytokines and chemokines, which regulate immune responses. This genetic upregulation enhances both innate and adaptive immunity, leading to better responses to bacterial, viral, and parasitic infections. In Asian seabass, for example, probiotics such as *Bacillus subtilis* have been shown to increase the expression of cytokines like interleukin-1 beta (IL-1β) and tumor necrosis factor-alpha (TNF-α), which play crucial roles in immune signaling during pathogen attacks.

Immune Response in Disease Prevention and Management

The immune-enhancing properties of probiotics and prebiotics make them valuable tools for disease prevention and management in aquaculture. Disease outbreaks are a major concern in fish farming, as they can lead to significant economic losses. Enhancing the immune function of farmed fish through dietary interventions with probiotics and prebiotics provides a sustainable and eco-friendly approach to reducing disease incidence without relying on antibiotics or other chemical treatments.

1. **Reduced Disease Severity**: Fish fed probiotic- or prebiotic-enriched diets have been shown to experience lower disease severity and faster recovery rates when exposed to pathogens. For example, studies on tilapia and carp infected with *Aeromonas hydrophila* have demonstrated that fish receiving probiotic-supplemented diets exhibit reduced

mortality and less severe clinical signs of disease. This reduced disease severity is attributed to the enhanced immune responses triggered by probiotics and prebiotics, which enable the fish to mount more effective defenses against invading pathogens.

2. **Increased Survival Rates**: The use of probiotics and prebiotics has been associated with higher survival rates in fish, especially during disease challenges. Enhanced immunity, along with better pathogen resistance and overall health, contributes to improved survival rates. This is particularly valuable in intensive aquaculture systems, where fish are more susceptible to infections due to high stocking densities and environmental stressors. For instance, studies on rainbow trout have shown that feeding with *Lactobacillus rhamnosus* or *Saccharomyces cerevisiae* increases survival during bacterial infections, making these probiotics effective in preventing disease outbreaks in commercial fish farming.

5. Current Research and Future Perspectives

The incorporation of probiotics and prebiotics into aquaculture has gained considerable attention due to their proven benefits for fish health, nutrition, and sustainability. However, despite significant advancements in the understanding of their roles, several challenges persist that require further exploration and innovation. Current research is addressing these gaps, while future studies will likely explore cutting-edge approaches to improve the efficacy and application of probiotics and prebiotics in aquaculture systems.

5.1 Strain-Specific Probiotics

One of the most promising areas of research focuses on the development of strain-specific probiotics, tailored to the unique physiological needs of different fish species and their production environments. While generic probiotics have shown benefits, it is clear that the effectiveness of a probiotic strain can vary widely depending on the fish species, water quality, temperature, and other environmental factors. Research is now aimed at isolating and identifying probiotic strains that are better adapted to the specific gut environment and immune system of target species such as tilapia, salmon, carp, and catfish.

For example, certain strains of Lactobacillus and Bacillus species have shown excellent performance in improving the gut health and immune function of warm-water fish like tilapia but may not perform as effectively in cold-water species like salmon. Tailoring probiotic strains to different fish species can optimize gut colonization, immune modulation, and growth performance, leading to more reliable outcomes in aquaculture. Additionally, multi-strain probiotics are being explored to create synergistic effects. These blends of complementary strains are designed to promote a more diverse and stable gut microbiota, improving both health and resilience to stress and disease.

5.2 Mechanistic Studies

A deeper understanding of the molecular mechanisms underlying the beneficial effects of probiotics and prebiotics is essential for advancing their application in aquaculture. Current research is focusing on elucidating how these functional feed additives interact with the fish gut microbiota, modulate immune responses, and influence nutrient metabolism at the cellular and molecular levels.

Some of the key areas being studied include:

- **Gene expression**: How probiotics and prebiotics influence the expression of genes involved in immune responses, gut integrity, and metabolic processes.
- **Microbiota dynamics**: Understanding how different prebiotics selectively stimulate beneficial bacteria and how probiotics alter the composition and metabolic activity of the gut microbiota.
- **Host-microbiota interactions**: Exploring the bidirectional communication between the gut microbiota and the fish's immune system, and how this interaction affects health and disease resistance.

These mechanistic studies are critical for optimizing the formulation of probiotic and prebiotic supplements to achieve desired health outcomes in fish. They also provide insights into how dietary interventions can be used to manage gut dysbiosis (microbial imbalance) and reduce the incidence of gut-related diseases in aquaculture systems.

5.3 Omics Technologies

The application of omics technologies—including genomics, transcriptomics, proteomics, and metabolomics—is revolutionizing the field of aquaculture by providing comprehensive insights into the molecular interactions between fish, their gut microbiota, and the feed they consume. These technologies are being used to explore the impact of probiotics and prebiotics on fish health and nutrition at an unprecedented level of detail.

- **Genomics** is helping researchers identify the specific genes responsible for the probiotic and prebiotic effects on gut health, immune function, and nutrient metabolism.
- **Transcriptomics** is used to study changes in gene expression in fish fed probiotic- and prebiotic-supplemented diets, shedding light on how these supplements affect key biological pathways.

- **Proteomics** and **metabolomics** are being employed to analyze the production of proteins and metabolites (such as short-chain fatty acids) by gut microbiota, and how these compounds influence fish health and growth.

By integrating data from these omics approaches, researchers can identify biomarkers of fish health and performance, enabling more targeted interventions to improve aquaculture productivity. Personalized nutrition based on the genetic and microbial profiles of fish could be the future of aquaculture, where probiotic and prebiotic supplements are customized to individual species, strains, or even farms.

5.4 Sustainability

As aquaculture grows, there is increasing pressure to minimize its environmental impact. The use of probiotics and prebiotics plays a key role in this effort by reducing reliance on antibiotics and other chemical treatments, which can contribute to the emergence of antibiotic-resistant pathogens and environmental pollution.

Current research is exploring ways to enhance the sustainability of probiotics and prebiotics in aquaculture:

- **Probiotic alternatives to antibiotics**: Probiotics are being developed as natural alternatives to antibiotics, reducing the need for chemical treatments in fish farming and promoting healthier, more resilient fish populations.
- **Novel sources of prebiotics**: Research is investigating new prebiotics derived from natural and sustainable sources, such as marine algae, plant extracts, and waste products from the food industry. These novel prebiotics may offer enhanced functionality and contribute to more sustainable feed production.
- **Reducing nutrient waste**: By improving nutrient digestion and absorption, probiotics and prebiotics reduce the excretion of undigested nutrients into the water, minimizing the environmental impact of aquaculture effluents on aquatic ecosystems.

As global demands for fish protein continue to rise, probiotics and prebiotics will be crucial for ensuring that aquaculture remains economically viable and environmentally sustainable.

6. Conclusion

The use of probiotics and prebiotics in aquaculture nutrition is rapidly transforming the industry by offering natural, effective solutions for improving gut health, immune function, and nutrient digestibility. These functional feed additives not only enhance fish growth performance and disease resistance but also contribute to more sustainable and eco-friendly aquaculture practices. As the body of research on probiotics and prebiotics continues to grow, it is becoming increasingly clear that these supplements have the potential to address many of the challenges facing modern fish farming, including the need for improved fish welfare and reduced reliance on antibiotics. Future research will focus on refining probiotic and prebiotic formulations to target specific fish species, production environments, and health goals. Omics technologies will play a critical role in this endeavor, providing a deeper understanding of the molecular interactions between fish and their gut microbiota. Additionally, the development of novel and sustainable prebiotic sources will further enhance the environmental benefits of aquaculture. As the industry embraces holistic and natural approaches to maintaining fish health, probiotics and prebiotics will continue to play a central role in the future of sustainable aquaculture, ensuring the production of healthy, high-quality seafood to meet the growing global demand.

Table: Probiotics, Prebiotics, and Synbiotics in Fish Gut Health Modulation

Aspect	Probiotics	Prebiotics	Synbiotics (Probiotics + Prebiotics)
Definition	Live microorganisms that confer health benefits by improving gut microbiota balance.	Non-digestible food components that selectively enhance beneficial gut microbes.	Combination of probiotics and prebiotics to provide synergistic benefits to gut health.
Common Types	Lactobacillus, Bifidobacterium, Bacillus, Saccharomyces	Oligosaccharides (FOS, MOS), inulin, and other polysaccharides	Paired probiotics and prebiotics for improved gut colonization and nutrient utilization.
Key Mechanisms of Action	- Competes with pathogens for nutrients and adhesion sites - Produces antimicrobial substances (bacteriocins, acids) - Boosts immune responses - Enhances nutrient digestion (enzyme production)	- Fermented by beneficial bacteria - Produces short-chain fatty acids (SCFAs) - Supports gut barrier integrity - Reduces inflammation	- Enhances survival and colonization of probiotics in the gut - Amplifies immune modulation and nutrient absorption

Effect on Gut Microbiota	Promotes beneficial bacteria and reduces pathogen colonization	Selectively stimulates growth of beneficial microbes (e.g., Lactobacillus, Bifidobacterium)	Boosts probiotic activity, improves gut microbial diversity, and stabilizes gut microbiota.
Impact on Immune Function	Stimulates innate and adaptive immune responses, increasing protective enzymes (e.g., lysozyme) and immune cell activity	Modulates immune function by promoting antibody production and gut immune cell activity	Stronger immune responses, reducing infection risk and enhancing pathogen resistance.
Contribution to Nutrient Absorption	Secretes enzymes (e.g., amylase, protease) for better breakdown of nutrients, improving feed conversion	Promotes SCFA production, enhancing nutrient absorption and gut health	Improved enzymatic activity and nutrient absorption, leading to optimized growth performance.
Growth Performance	Improves feed conversion ratio (FCR), weight gain, and protein efficiency	Enhances overall growth by improving nutrient uptake and gut health	Maximizes fish growth rates, FCR, and overall productivity through synergistic effects.
Current Research Focus	- Strain-specific probiotics for different species - Understanding probiotic-host interactions	- Novel prebiotics from algae, plants, and marine organisms - Prebiotics' role in sustainability	- Optimizing synbiotic formulations for various fish species - Reducing antibiotic use through synbiotics
Future Directions	- Development of next-gen probiotics (genetically engineered strains) - Multi-strain probiotics tailored to aquaculture systems	- Discovery of new, eco-friendly prebiotics for sustainable aquaculture	Expanding synbiotics for more species and environments, contributing to antibiotic-free and sustainable aquaculture practices.

References

Akhter, N., Wu, B., Memon, A. M., & Mohsin, M. (2015). Probiotics and prebiotics associated with aquaculture: A review. Fish & Shellfish Immunology, 45(2), 733-741.

Balcázar, J. L., de Blas, I., Ruiz-Zarzuela, I., Cunningham, D., Vendrell, D., & Múzquiz, J. L. (2006). The role of probiotics in aquaculture. Veterinary Microbiology, 114(3-4), 173-186.

Burr, G., Gatlin, D. M., & Ricke, S. C. (2005). Microbial ecology of the gastrointestinal tract of fish and the potential application of prebiotics and probiotics in finfish aquaculture. Journal of the World Aquaculture Society, 36(4), 425-436.

Carnevali, O., de Vivo, L., Sulpizio, R., Gioacchini, G., Olivotto, I., Silvi, S., & Cresci, A. (2006). Growth improvement by probiotic in European sea bass juveniles (Dicentrarchus labrax, L.), with particular attention to IGF-1, myostatin, and cortisol gene expression. Aquaculture, 258(1-4), 430-438.

Dimitroglou, A., Merrifield, D. L., Carnevali, O., Picchietti, S., Avella, M., Daniels, C., & Davies, S. J. (2011). Microbial manipulations to improve fish health and production – A Mediterranean perspective. Fish & Shellfish Immunology, 30(1), 1-16.

Doan, H. V., Hoseinifar, S. H., Naraballobh, W., & Jaturasitha, S. (2020). Host-associated probiotics: A key factor in sustainable aquaculture. Aquaculture Research, 51(3), 1236-1249.

Gatesoupe, F. J. (1999). The use of probiotics in aquaculture. Aquaculture, 180(1-2), 147-165.

Gomez-Gil, B., Roque, A., & Turnbull, J. F. (2000). The use and selection of probiotic bacteria for use in the culture of larval aquatic organisms. Aquaculture, 191(1-3), 259-270.

Hai, N. V. (2015). The use of probiotics in aquaculture. Journal of Applied Microbiology, 119(4), 917-935.

He, S., Zhou, Z., Liu, Y., Shi, P., Yao, B., Ringø, E., & Yoon, I. (2009). Effects of dietary Saccharomyces cerevisiae fermentation product on growth performance, intestinal autochthonous bacterial communities, gut morphology, and innate immunity of hybrid tilapia (Oreochromis niloticus ♀ × Oreochromis aureus ♂) cultured in cages. Aquaculture, 294(1-2), 99-107.

Merrifield, D. L., & Ringø, E. (Eds.). (2014). Aquaculture Nutrition: Gut Health, Probiotics, and Prebiotics. Wiley-Blackwell. ISBN: 978-1-118-91514-1

Merrifield, D. L., Dimitroglou, A., Foey, A., Davies, S. J., Baker, R. T. M., Bøgwald, J., & Ringø, E. (2010). The current status and future focus of probiotic and prebiotic applications for salmonids. Aquaculture, 302(1-2), 1-18.

Nayak, S. K. (2010). Role of gastrointestinal microbiota in fish. Aquaculture Research, 41(11), 1553-1573.

Nikoskelainen, S., Ouwehand, A., Bylund, G., Salminen, S., & Lilius, E. M. (2001). Immune enhancement in rainbow trout (Oncorhynchus mykiss) by potential probiotic bacteria (Lactobacillus rhamnosus). Aquaculture, 198(3-4), 275-286.

Pandiyan, P., Balaraman, D., Thirunavukkarasu, R., George, E. G. J., Subaramaniyan, K., Manikkam, S., & Sadayappan, B. (2013). Probiotics in aquaculture. Drug Invention Today, 5(1), 55-59.

Peng, M., Liu, D., Liu, H., Zhou, Y., & Li, P. (2019). The effect of prebiotics on gut microbiota in tilapia (Oreochromis niloticus) and their resistance to Aeromonas hydrophila infection. Fish & Shellfish Immunology, 91, 138-147.

Rangel, A. H. N., do Nascimento, L. C., da Silva, J. G., de Macedo, F. M., & Urbano, S. A. (2017). Use of probiotics in aquatic organisms: Review. African Journal of Agricultural Research, 12(26), 2228-2235.

Ringø, E., & Birkbeck, T. H. (1999). Intestinal microflora of fish larvae and fry. Aquaculture Research, 30(2), 73-93. https://doi.org/10.1046/j.1365-2109.1999.00302.x

Ringø, E., & Song, S. K. (2016). Application of dietary supplements (synbiotics and probiotics in combination with prebiotics) in aquaculture. Aquaculture Nutrition, 22(1), 4-24.

Ringø, E., Olsen, R. E., Vecino, J. L. G., Wadsworth, S., & Song, S. K. (2012). Use of immunostimulants and nucleotides in aquaculture: A review. Journal of Marine Science: Research & Development, 2(1), 2-8.

Song, S. K., Beck, B. R., Kim, D., Park, J., Kim, J., & Ringø, E. (2014). Prebiotics as immunostimulants in aquaculture: A review. Fish & Shellfish Immunology, 40(1), 40-48.

Tovar-Ramírez, D., Mazurais, D., Gatesoupe, F. J., Quazuguel, P., Cahu, C. L., & Zambonino-Infante, J. L. (2010). Dietary probiotic live yeast modulates antioxidant enzyme activities and gene expression of sea bass larvae challenged with Photobacterium damselae subsp. piscicida. Aquaculture, 300(1-4), 142-147.

Vine, N. G., Leukes, W. D., & Kaiser, H. (2006). Probiotics in marine larviculture. FEMS Microbiology Reviews, 30(3), 404-427.

Wu, Y., Zhou, Y., Lu, X., Song, W., & Mai, K. (2013). The effect of dietary fructooligosaccharide on growth, non-specific immunity, intestinal enzyme activities, and morphology of Japanese seabass, Lateolabrax japonicus. Fish & Shellfish Immunology, 35(1), 56-62.

Zhou, Z., Ringø, E., Olsen, R. E., & Song, S. K. (2018). Dietary effects of soybean products on gut microbiota and immunity of aquatic animals: A review. Aquaculture Nutrition, 24(1), 644-665.

7

Insect-Based Protein in Aquafeeds A Sustainable Alternative

1. Introduction

Aquaculture, the farming of fish and other aquatic organisms, has experienced exponential growth over the last few decades, becoming one of the fastest-growing food production sectors globally. This expansion is driven by the rising global demand for seafood, fueled by population growth, increased per capita consumption, and the recognition of fish as a healthy and nutrient-rich food source. According to the Food and Agriculture Organization (FAO), aquaculture now supplies more than half of the world's fish for human consumption, a significant leap from the 1980s when it contributed only a fraction of global seafood production. However, this rapid growth poses challenges, particularly with respect to sustainable feed production.

One of the primary concerns in aquaculture is the reliance on fishmeal, a high-protein feed ingredient traditionally sourced from small, wild-caught fish species such as anchovies, sardines, and menhaden. Fishmeal is rich in essential nutrients, particularly amino acids, omega-3 fatty acids, and micronutrients, making it an ideal component for aquafeeds. However, its dependence on wild fish stocks has raised sustainability concerns, as the continuous harvesting of these species can lead to overfishing, biodiversity loss, and ecosystem degradation. Moreover, the rising demand for fishmeal in aquafeeds, coupled with finite fishery resources, has driven up its price, making it less economically viable for aquaculture operations, especially for small-scale farmers.

In this context, the search for alternative, sustainable, and cost-effective protein sources for aquafeeds has become a pressing issue. Plant-based proteins such as soybean meal have been explored, but they come with their own set of challenges, including anti-nutritional factors, limited essential amino acid profiles, and competition with human food production. As the aquaculture industry seeks to reduce its environmental footprint while maintaining high feed quality, insect-based protein has emerged as a promising solution.

Insects are a natural part of the diet for many wild fish species, making them biologically relevant feed components for aquaculture. Insects are not only rich in high-quality protein but also provide essential amino acids, fatty acids, vitamins, and minerals that are crucial for fish growth and health. Black soldier fly larvae (BSFL), mealworms, and crickets are among the most studied insect species for use in animal feed, and they have shown great potential as sustainable alternatives to fishmeal. The amino acid profile of insect-based proteins is comparable to fishmeal, making them suitable for replacing or supplementing traditional feed ingredients in aquafeeds.

Beyond their nutritional value, insects offer significant environmental advantages. One of the most compelling aspects of insect farming is its ability to utilize organic waste streams, such as food scraps, agricultural by-products, and even manure, as feed for insects. This capacity for upcycling waste into high-quality protein not only reduces the need for traditional agricultural inputs but also contributes to waste reduction and the circular economy. Insects can be produced on a large scale with minimal environmental impact compared to conventional livestock and fishmeal production. They require less land, water, and energy, and they generate fewer greenhouse gas emissions, making them a highly sustainable option for future feed production.

This chapter explores the potential of insect-based protein as a sustainable alternative for aquafeeds in detail. It examines the nutritional value of insects, their environmental benefits, and the challenges associated with scaling up insect farming for the global aquaculture industry. Furthermore, it discusses the future prospects for insect-based protein, highlighting ongoing research and innovations that aim to make insect protein a mainstream component of aquafeeds. By addressing the need for sustainable feed solutions, insect-based protein offers a promising path forward for aquaculture, helping to reduce the industry's reliance on finite marine resources while supporting global food security.

2. Nutritional Value of Insect-Based Protein

The nutritional profile of insect-based proteins makes them a valuable and sustainable alternative to traditional protein sources such as fishmeal in aquafeeds. Several insect species have been extensively studied for their potential use in aquaculture, with black soldier fly larvae (*Hermetia illucens*), mealworms (*Tenebrio molitor*), and crickets (Gryllidae spp.) emerging as the most promising candidates. These insects offer a wide array of essential nutrients, including high-quality protein, essential fatty acids, vitamins, minerals, and bioactive compounds that support fish health, growth, and performance.

2.1 High-Quality Protein and Amino Acid Profile

Insect-based proteins are particularly attractive due to their high protein content, which typically ranges between 40% and 60% of their dry weight, depending on the species and rearing conditions. This is comparable to fishmeal, which contains around 60-70% protein. More importantly, the amino acid profile of insect protein is well-balanced and highly suitable for use in aquafeeds.

Essential amino acids such as lysine and methionine, which are often limiting in plant-based feed ingredients like soybean meal, are present in significant quantities in insect proteins. Lysine and methionine are critical for fish growth, muscle development, and metabolic functions. The presence of these amino acids in sufficient amounts makes insect protein a valuable alternative to fishmeal, addressing the nutritional deficiencies often associated with plant-based feeds. For example, black soldier fly larvae (BSFL) protein contains approximately 5.8% lysine and 2.3% methionine, which closely matches the amino acid requirements of many fish species, including carnivorous species like salmon and trout. The amino acid composition of insect protein varies slightly depending on the insect species, but in general, insects offer a protein source that is highly digestible and nutritionally adequate for most farmed fish species. Studies have shown that the growth performance of fish fed insect-based diets is comparable to those fed traditional fishmeal-based diets, demonstrating the viability of insects as a feed ingredient.

2.2 Essential Fatty Acids and Lipid Profile

In addition to protein, insects provide an excellent source of essential fatty acids, which are important for fish health, growth, and reproduction. Black soldier fly larvae, for example, are particularly rich in beneficial lipids, including both omega-3 and omega-6 fatty acids. While the content of omega-3 fatty acids in insect proteins may not reach the same levels as in fishmeal, they still offer a balanced lipid profile that can contribute to fish health.

Omega-3 fatty acids, such as eicosapentaenoic acid (EPA) and docosahexaenoic acid (DHA), are critical for fish, especially carnivorous species, as they play vital roles in brain function, immune response, and inflammation control. Although insect species do not produce large amounts of long-chain omega-3 fatty acids like EPA and DHA, they can be supplemented with algae or other marine-based ingredients to balance the lipid profile of aquafeeds. Additionally, insects contain high levels of medium-chain fatty acids, particularly lauric acid, which has antimicrobial properties that can support fish health by inhibiting pathogenic bacteria in the gut.

2.3 Bioactive Compounds: Antimicrobial Peptides and Chitin

Insects also offer unique bioactive compounds that contribute to fish health beyond basic nutrition. Two key components are antimicrobial peptides (AMPs) and chitin, both of which have functional roles in improving fish immunity and gut health.

- **Antimicrobial Peptides (AMPs)**: Insects naturally produce AMPs as part of their immune defense system against pathogens. These peptides have been shown to inhibit the growth of harmful bacteria, fungi, and viruses. When included in aquafeeds, AMPs can enhance the fish's immune response and offer protection against diseases. For instance, studies have demonstrated that fish fed diets containing insect protein exhibit lower mortality rates when exposed to bacterial infections such as *Aeromonas hydrophila* or *Vibrio anguillarum.*
- **Chitin and Chitosan**: Chitin is a polysaccharide found in the exoskeletons of insects. It has prebiotic properties, meaning it can selectively stimulate the growth of beneficial gut bacteria in fish. When chitin is broken down by gut microbes into its derivative, chitosan, it further enhances gut health by supporting the gut barrier, reducing intestinal inflammation, and boosting immune function. Chitin also contributes to improved digestion by promoting enzymatic activity and nutrient absorption, thus enhancing overall feed efficiency.

2.4 Digestibility and Growth Performance

The digestibility of insect-based proteins is another critical factor determining their suitability for aquafeeds. Numerous studies have demonstrated that the digestibility of insect protein can be comparable to or even higher than that of fishmeal, depending on the species of insect and the processing methods used. For example, black soldier fly larvae meal has been reported to have a digestibility coefficient similar to fishmeal, making it an efficient and nutritionally valuable feed ingredient. Processing techniques such as drying, grinding, and defatting can influence the nutritional quality and digestibility of insect protein. Proper processing helps to reduce anti-nutritional factors and optimize the nutrient content, ensuring that the protein and other essential nutrients are readily available for digestion and absorption by fish. Moreover, insect protein has been shown to improve feed conversion ratios (FCR) in some species, meaning fish can convert feed into body mass more efficiently when fed diets containing insect protein.

In aquaculture trials, insect-based protein has been shown to promote growth performance comparable to that of fishmeal-based diets. For instance, several studies have reported that fish species such as tilapia, rainbow trout, and European sea bass fed diets containing black soldier fly larvae meal exhibit similar or even superior growth rates compared to fishmeal-fed counterparts. This makes insect protein a promising alternative for maintaining high productivity in aquaculture while reducing dependence on marine resources.

2.5 Mineral and Vitamin Content

Insects are also a rich source of essential minerals and vitamins, including calcium, phosphorus, magnesium, zinc, iron, and B-vitamins. These micronutrients are important for various physiological functions in fish, including bone development, metabolic processes, and immune responses. The mineral content of insect-based proteins can vary depending on the insect species and their diet, but in general, they provide sufficient quantities to support the nutritional needs of farmed fish. For instance, black soldier fly larvae are known to have high levels of calcium and phosphorus, which are critical for fish skeletal development and overall health. The availability of these minerals in insect protein can help reduce the need for additional mineral supplementation in aquafeeds, further improving the cost-effectiveness of insect-based diets.

3. Environmental Benefits of Insect Protein

The use of insect-based protein in aquafeeds offers several notable environmental advantages, making it a sustainable alternative to traditional protein sources such as fishmeal and soybean meal. As the aquaculture industry continues to grow, there is an increasing need for feed ingredients that can meet the nutritional needs of fish without contributing to ecological degradation. Insect farming provides a promising solution due to its resource-efficient nature, minimal environmental impact, and ability to contribute to a circular economy.

3.1 Reduced Reliance on Wild Fish Stocks

One of the most significant environmental benefits of using insect protein in aquafeeds is the reduction in dependence on wild fish stocks. Fishmeal, traditionally the primary protein source in aquaculture feeds, is derived from wild-caught fish such as anchovies, sardines, and mackerel. The increasing demand for fishmeal has placed substantial pressure on wild fish populations, contributing to overfishing and the depletion of marine ecosystems. This unsustainable practice threatens biodiversity and the long-term viability of both fisheries and aquaculture. In contrast, insect protein can be produced without impacting wild fish stocks, offering a more sustainable alternative. Insects are a natural part of the diet of many fish species, and their use in aquafeeds aligns with the biological and ecological needs of farmed fish. By replacing fishmeal with insect protein, the aquaculture industry can help alleviate pressure on marine ecosystems and contribute to the conservation of biodiversity.

3.2 Efficient Use of Resources

Insect farming is highly resource-efficient compared to traditional livestock and crop-based protein sources. Insects require significantly less land, water, and feed than conventional protein sources like soybean meal or animal agriculture. For example, black soldier fly larvae (BSFL) can be raised on a variety of organic waste materials, including food scraps, agricultural by-products, and manure. This ability to upcycle low-value organic waste into high-quality protein reduces the need for arable land and water resources, making insect farming a sustainable solution for feed production.

- **Land Use**: Insect farming requires minimal land area compared to traditional crop and livestock farming. For instance, producing 1 kilogram of insect protein requires much less land than producing the same amount of protein from soybeans or poultry. This reduced land footprint helps prevent deforestation and habitat loss, which are major concerns associated with agricultural expansion.
- **Water Use**: Insects also require far less water than conventional protein sources. For example, producing 1 kilogram of insect protein requires only a fraction of the water needed to produce the same amount of protein from beef, poultry, or soybeans. Given the increasing scarcity of freshwater resources, this reduced water demand is a significant environmental advantage.
- **Feed Conversion Efficiency**: Insects are highly efficient at converting feed into biomass. Black soldier fly larvae, for instance, can convert organic waste into protein-rich biomass within a matter of days, achieving a feed conversion ratio (FCR) of less than 2:1. This means that for every 2 kilograms of feed, 1 kilogram of insect biomass is produced. In comparison, fishmeal production and traditional livestock farming have much higher FCRs, requiring significantly more feed inputs to produce the same amount of protein.

3.3 Minimal Greenhouse Gas Emissions

Another environmental benefit of insect farming is its low greenhouse gas (GHG) emissions. Conventional livestock farming, particularly cattle and sheep, is a major contributor to GHG emissions, particularly methane (CH□) and nitrous oxide (N□O), which have a high global warming potential. In contrast, insects produce negligible amounts of GHGs during their life cycle.

The carbon footprint of insect farming is significantly lower than that of traditional protein production methods, making it an environmentally friendly option for aquafeeds. For example, studies have shown that black soldier fly larvae farming produces significantly lower emissions of carbon dioxide (CO□), methane, and nitrous oxide compared to beef, pork, and poultry production. The low emission levels associated with insect farming contribute to the mitigation of climate change and the promotion of more sustainable agricultural practices.

3.4 Waste Management and Circular Economy

One of the most compelling environmental benefits of insect protein production is its role in waste management and the promotion of a circular economy. Insects, particularly black soldier fly larvae, can be reared on a variety of organic waste materials, including food waste, agricultural residues, and animal manure. This ability to convert low-value waste into high-quality protein is a key factor in the sustainability of insect farming. Food waste is a significant global issue, with millions of tons of edible food being discarded each year. Insect farming offers a practical solution to this problem by transforming food waste into a valuable resource for aquaculture feeds. By upcycling organic waste into insect protein, the industry can reduce the volume of waste sent to landfills, lower methane emissions from decomposing waste, and create a closed-loop system that promotes resource efficiency. Moreover, the by-products of insect farming, such as insect frass (excrement), can be used as organic fertilizer, further enhancing the sustainability of the production process. This contributes to a circular economy where waste products are continuously reused, reducing the need for synthetic fertilizers and minimizing environmental pollution.

3.5 Preservation of Biodiversity

The use of insect-based protein in aquafeeds can also contribute to the preservation of biodiversity by reducing the environmental impact of feed production. Conventional feed ingredients, such as soybeans and fishmeal, are associated with significant environmental degradation, including deforestation, habitat destruction, and loss of biodiversity. Soybean cultivation, in particular, has been linked to deforestation in regions like the Amazon rainforest, where large areas of land are cleared for agricultural expansion. By adopting insect protein as a sustainable alternative, the aquaculture industry can reduce its reliance on soybeans and other land-intensive crops, helping to preserve ecosystems and protect biodiversity. Additionally, reducing the demand for fishmeal can help alleviate pressure on marine ecosystems, promoting the recovery of fish populations and supporting the long-term sustainability of global fisheries.

3.6 Localized and Scalable Production

Insect farming can be easily scaled and localized, further contributing to its environmental benefits. Insects can be reared in controlled environments with minimal space requirements, making them suitable for both small-scale and industrial-scale production. This flexibility allows insect farming to be adapted to local conditions, reducing the need for long-distance transportation of feed ingredients and lowering the associated carbon emissions. In addition, insect farming can be integrated into existing agricultural systems, where organic waste from crop and livestock production can be used as feed for insects. This integration not only enhances resource efficiency but also provides a sustainable source of protein that can be produced locally, reducing the environmental impact of feed production and distribution.

Table: Key Aspects of Insect-Based Protein in Aquafeeds: A Sustainable Alternative

Aspect	Details
Insect Species	- Black Soldier Fly Larvae (*Hermetia illucens*)
	- Mealworms (*Tenebrio molitor*)
	- Crickets (Gryllidae spp.)
Nutritional Composition	**Protein**: High protein content with a well-balanced amino acid profile, comparable to fishmeal.
	Fatty Acids: Contains essential fatty acids, including omega-3 and omega-6.
	Chitin: Found in exoskeletons, with prebiotic properties supporting gut health.
	Antimicrobial Peptides (AMPs): Present in some insect species, improving fish immune response.
	Digestibility: Comparable to fishmeal depending on processing methods used.
Environmental Benefits	**Low Resource Use**: Insects can be farmed on organic waste, reducing the need for land and water resources.
	Feed Conversion Efficiency: Insects efficiently convert low-value waste into high-quality protein within days.
	Reduced GHG Emissions: Insect farming generates lower greenhouse gas emissions compared to traditional livestock.
	Waste Reduction: Insects upcycle agricultural and food waste into valuable feed ingredients, supporting a circular economy.
Challenges	**Regulatory Approval**: Not all insect species are yet approved for use in aquafeeds worldwide.
	Cost: Initial production costs of insect protein remain higher than traditional protein sources like fishmeal.
	Consumer Perception: Public acceptance of using insect-based feeds in seafood production is still growing.

Technological Advancements	**AI and Automation**: Use of artificial intelligence and automated farming systems to improve production efficiency.
	Breeding Programs: Selective breeding of insects to enhance growth rates, nutritional content, and adaptability to different waste streams.
	Processing Methods: New technologies to improve the digestibility and safety of insect-based feed ingredients.
Economic Impact	**Cost Reduction**: Scaling up production through advanced farming methods is expected to lower costs over time.
	Resource Efficiency: Insect farming contributes to resource-efficient feed production by utilizing waste streams.
	Job Creation: Expansion of the insect farming industry may create new employment opportunities in the agri-tech sector.
Sustainability Contributions	**Reduced Reliance on Fishmeal:** Insect protein offers an alternative to marine-sourced fishmeal, easing pressure on overexploited fish stocks.
	Circular Economy Integration: Insects fit into circular economy models by converting waste into valuable protein, reducing the environmental impact of food production.
Future Prospects	- Development of next-generation farming technologies to increase scalability and reduce costs.
	- Greater regulatory approvals for a wider range of insect species in aquafeeds.
	- Increased public awareness and consumer acceptance of insect-based proteins as sustainable solutions for aquaculture.

6. Future Prospects

The future of insect-based protein in aquafeeds looks highly promising, with increasing interest from both the aquaculture industry and the scientific community. Several factors, including advancements in insect farming technologies, regulatory developments, consumer awareness, and ongoing research, are driving the growth and viability of insect protein as a sustainable feed ingredient for aquaculture.

6.1 Advancements in Farming Technologies

The use of advanced technologies such as artificial intelligence (AI), robotics, and automation is expected to revolutionize insect farming, making it more efficient, scalable, and cost-effective. AI-driven systems can optimize key aspects of insect farming, including feed management, environmental control, and monitoring of insect growth and health. Automation will play a vital role in streamlining the harvesting, processing, and packaging of insect biomass, reducing labor costs and enhancing the overall efficiency of insect production systems. These technological advancements will also allow for large-scale production of insect protein, making it more accessible and affordable for the aquaculture industry. As production costs decrease and yields increase, insect-based protein will become a more attractive alternative to conventional protein sources like fishmeal and soybean meal. This scalability will help meet the growing demand for sustainable aquafeeds, particularly as the global population continues to rise and pressure on traditional protein sources intensifies.

6.2 Insect Breeding Programs and Genetic Improvement

The development of insect-breeding programs aimed at improving key traits such as growth rates, nutritional content, and adaptability to various organic waste streams will further enhance the viability of insect protein for aquafeeds. Selective breeding techniques, similar to those used in conventional livestock and aquaculture breeding programs, can be applied to insect species to increase protein yield, enhance feed conversion efficiency, and improve their overall nutritional profile. For example, breeding programs could focus on increasing the levels of essential amino acids, fatty acids, and beneficial bioactive compounds such as antimicrobial peptides (AMPs) in insect protein. Additionally, breeding insects that are better suited to specific environmental conditions or waste substrates would increase the flexibility and sustainability of insect farming operations. Such advancements would allow the aquaculture industry to tailor insect-based feeds to meet the nutritional needs of different fish species more effectively.

6.3 Research on Nutritional and Functional Benefits

Ongoing research is exploring the nutritional and functional benefits of insect-based protein in aquaculture. While many studies have already demonstrated the comparable nutritional value of insect protein to fishmeal, there is growing interest in understanding the additional health benefits that insects may offer. For instance, insect protein contains bioactive compounds like chitin, lauric acid, and antimicrobial peptides that can enhance fish gut health, immune function, and disease resistance. Further research into the role of these compounds in fish nutrition could lead to the development of functional feeds that not only provide essential nutrients but also improve fish welfare and reduce the need for antibiotics and other chemical treatments. This aligns with the global trend toward more natural and sustainable approaches to aquaculture, where feed ingredients play a crucial role in maintaining fish health and productivity.

6.4 Regulatory Developments

As the use of insect-based protein in aquafeeds gains momentum, regulatory frameworks are evolving to support its commercialization. Several countries and regions, including the European Union and the United States, have already approved certain insect species, such as black soldier fly larvae (Hermetia illucens), for use in animal feeds, including aquafeeds. However, further regulatory approvals are needed for additional insect species and applications to ensure that insect-based protein can be widely adopted across the global aquaculture industry. Regulatory bodies are also focusing on setting safety and quality standards for insect farming, processing, and feed formulation. These standards are essential for ensuring that insect protein is produced in a safe, hygienic, and environmentally responsible manner. As more regulatory approvals are granted, the market for insect-based protein is expected to expand, opening up new opportunities for innovation and growth in the aquafeed sector.

6.5 Consumer Perception and Market Acceptance

Consumer perception of insect-based protein in aquafeeds is another important factor that will influence its future prospects. While insect protein is widely accepted as a feed ingredient for animals, there is still some resistance to the idea of using insects as a source of nutrition for human-consumed animals, such as fish. Educating consumers about the environmental and nutritional benefits of insect protein, as well as its safety and efficacy, will be crucial in overcoming this resistance. Public awareness campaigns, transparency in sourcing, and clear labeling of products that contain insect-based ingredients can help build consumer trust and acceptance. As sustainability becomes a more prominent concern for both consumers and industry stakeholders, the environmental benefits of insect protein will likely drive increased demand for aquaculture products that use eco-friendly feed ingredients.

6.6 Integration into Circular Economy and Sustainability Goals

Insect-based protein has the potential to play a significant role in achieving broader sustainability goals and supporting the development of a circular economy in aquaculture. Insects can be reared on organic waste, such as food scraps, agricultural by-products, and animal manure, transforming low-value waste into high-quality protein. This upcycling of waste materials not only reduces the environmental impact of insect farming but also contributes to more sustainable waste management practices. In the future, insect farming could be integrated into larger agricultural systems, where organic waste from food production and livestock farming is used to produce insect protein, and the by-products of insect farming (such as frass, or insect excrement) are used as organic fertilizers. This closed-loop system would minimize waste, promote resource efficiency, and reduce the overall environmental footprint of food production.

6.7 Reducing Reliance on Marine Resources

One of the most pressing challenges facing the aquaculture industry is its reliance on marine resources, particularly fishmeal, which is derived from wild-caught fish. As global fish stocks continue to decline due to overfishing and environmental degradation, the need for alternative protein sources has become increasingly urgent. Insect-based protein offers a sustainable solution to this challenge by providing a biologically appropriate feed ingredient that does not rely on marine resources. By reducing the industry's dependence on fishmeal, insect-based protein can help alleviate pressure on wild fish populations, promote biodiversity conservation, and support the long-term sustainability of global fisheries. As the demand for seafood continues to rise, adopting insect-based protein in aquafeeds will be critical to ensuring the responsible and sustainable growth of the aquaculture industry.

7. Conclusion

Insect-based protein represents a sustainable and innovative alternative to traditional feed ingredients in aquaculture. Its high nutritional value, low environmental footprint, and ability to reduce reliance on fishmeal make it a key solution to the challenges facing modern aquaculture. While there are still hurdles to overcome, such as regulatory approval, production costs, and consumer acceptance, ongoing research and technological advancements are likely to address these challenges and unlock the full potential of insect-based protein. As the global aquaculture industry continues to expand, the adoption of insect protein in aquafeeds will play an increasingly important role in ensuring the sector's sustainability, reducing its environmental impact, and meeting the growing demand for seafood. By integrating insect protein into feed formulations, the aquaculture industry can contribute to a more resource-efficient and environmentally responsible food production system, supporting global efforts to promote sustainable development and food security.

References

Barroso, F. G., de Haro, C., Sánchez-Muros, M. J., Venegas, E., Martínez-Sánchez, A., & Pérez-Bañón, C. (2014). The potential of various insect species for use as food for fish. Aquaculture, 422, 193-201.

Belghit, I., Liland, N. S., Gjesdal, P., Biancarosa, I., Menchetti, E., Li, Y., ... & Lock, E. J. (2019). Black soldier fly larvae meal can replace fish meal in diets of sea-water phase Atlantic salmon (Salmo salar). Aquaculture, 503, 609-619.

Cummins, V. C., Rawles, S. D., Thompson, K. R., Velasquez, A., Kobler, C., & Hager, J. (2017). Evaluation of black soldier fly (Hermetia illucens) larvae as a replacement fishmeal protein ingredient in practical diets for pacific white shrimp (Litopenaeus vannamei). Aquaculture, 473, 337-344.

Devic, E., Leschen, W., Murray, F., & Little, D. C. (2018). Growth performance, feed utilization and body composition of European sea bass (Dicentrarchus labrax) fed diets containing black soldier fly (Hermetia illucens) larvae meal. Aquaculture Nutrition, 24(1), 416-423.

Gasco, L., Biancarosa, I., & Liland, N. S. (2020). From waste to feed: A review of recent knowledge on insects as producers of protein and fat for animal feeds. Current Opinion in Green and Sustainable Chemistry, 23, 67-79.

Henry, M., Gasco, L., Piccolo, G., & Fountoulaki, E. (2015). Review on the use of insects in the diet of farmed fish: Past and future. Animal Feed Science and Technology, 203, 1-22.

Józefiak, A., Józefiak, D., Kierończyk, B., Rawski, M., Mazurkiewicz, J., & Sugiharto, S. (2016). Insects—a natural nutrient source for poultry—A review. Annals of Animal Science, 16(2), 297-313.

Kroeckel, S., Harjes, A. G. E., Roth, I., Katz, H., Wuertz, S., Susenbeth, A., & Schulz, C. (2012). When a turbot catches a fly: Evaluation of a pre-pupae meal of the black soldier fly (Hermetia illucens) as fishmeal substitute–Growth performance and chitin degradation in juvenile turbot (Psetta maxima). Aquaculture, 364, 345-352.

Li, Q., Zheng, L., Qiu, N., Cai, H., Tomberlin, J. K., Yu, Z., & Zhang, J. (2011). Bioconversion of dairy manure by black soldier fly (Diptera: Stratiomyidae) for biodiesel and sugar production. Waste Management, 31(6), 1316-1320.

Lock, E. R., Arnesen, R., Ringo, E., & Birkbeck, T. H. (2020). Advances in insect protein utilization in aquafeeds: Impact on growth performance, nutrient digestibility, and fish health. Aquaculture Research, 51(1), 7-22.

Makkar, H. P., Tran, G., Heuzé, V., & Ankers, P. (2014). State-of-the-art on use of insects as animal feed. Animal Feed Science and Technology, 197, 1-33.

Newton, L., Sheppard, C., Watson, D. W., Burtle, G., & Dove, R. (2005). Using the black soldier fly, Hermetia illucens, as a value-added tool for the management of swine manure. Journal of Animal and Veterinary Advances, 4(12), 280-285.

Nogales-Mérida, S., Gobbi, P., Jozefiak, D., Mazurkiewicz, J., Dudek, K., Rawski, M., ... & Józefiak, A. (2019). Insect meals in fish nutrition. Reviews in Aquaculture, 11(4), 1080-1103.

Rumpold, B. A., & Schlüter, O. K. (2013). Nutritional composition and safety aspects of edible insects. Molecular Nutrition & Food Research, 57(5), 802-823.

Sánchez-Muros, M. J., Barroso, F. G., & Manzano-Agugliaro, F. (2014). Insect meal as renewable source of food for animal feeding: A review. Journal of Cleaner Production, 65, 16-27.

Sealey, W. M., Gaylord, T. G., Barrows, F. T., Tomberlin, J. K., McGuire, M. A., Ross, C., & St-Hilaire, S. (2011). Sensory analysis of rainbow trout, Oncorhynchus mykiss, fed enriched black soldier fly prepupae, Hermetia illucens. Journal of the World Aquaculture Society, 42(1), 34-45.

Tacon, A. G. J., & Metian, M. (2008). Global overview on the use of fish meal and fish oil in industrially compounded aquafeeds: Trends and future prospects. Aquaculture, 285(1-4), 146-158.

Tschirner, M., & Simon, A. (2015). Influence of insect meal on broiler performance and health status: A feeding experiment. Berichte über Landwirtschaft, 93(1), 1-7.

van Huis, A. (2013). Potential of insects as food and feed in assuring food security. Annual Review of Entomology, 58, 563-583.

Veldkamp, T., & Bosch, G. (2015). Insects: A protein-rich feed ingredient in pig and poultry diets. Animal Frontiers, 5(2), 45-50.

8

Algal Biomass
A Superfood for Aquaculture

1. Introduction

The global demand for seafood has surged in recent decades, driven by population growth, rising incomes, and an increasing awareness of the health benefits associated with seafood consumption. In response, the aquaculture industry has expanded rapidly, making it one of the fastest-growing food production sectors in the world. As aquaculture continues to scale up, the need for sustainable and nutrient-rich feed ingredients has become a central concern. Traditional feed ingredients, such as fishmeal and soy, have significant environmental and ethical challenges. Fishmeal, derived from wild-caught fish, contributes to the depletion of marine resources, while soy cultivation is associated with deforestation and high water usage.

In this context, algal biomass has emerged as a promising, sustainable alternative to traditional aquafeed ingredients. Often referred to as a "superfood," algae, which include both microalgae and macroalgae (seaweeds), offer a rich source of essential nutrients that support the growth, health, and productivity of farmed aquatic species. Algal biomass is increasingly recognized for its unique ability to provide a wide range of proteins, essential fatty acids (such as omega-3 and omega-6), vitamins, minerals, and bioactive compounds that are critical to the optimal functioning of fish and other aquatic organisms.

What makes algal biomass particularly attractive as an aquafeed ingredient is its sustainability. Unlike land-based crops or marine-based fishmeal, algae can be cultivated in various environments with minimal ecological impact. They require limited land, freshwater, or chemical inputs, and many species of algae can be grown in saline or brackish water, areas unsuitable for conventional agriculture. Algae also exhibit fast growth rates, and their cultivation can be integrated with waste recycling systems, enhancing their role in circular economies. This characteristic makes algae not only a nutritionally valuable resource but also an environmentally responsible choice for the aquaculture industry.

The Role of Algal Biomass in Aquaculture

The incorporation of algal biomass in aquafeeds addresses several key challenges faced by the aquaculture industry. Algal biomass contributes to:

1. **Nutritional Enhancement**: The diverse nutrient profile of algae, including essential amino acids, fatty acids, and antioxidants, makes it an ideal supplement to improve the quality of aquafeeds.
2. **Reducing Fishmeal Dependency**: The declining availability and rising cost of fishmeal have made it unsustainable for long-term use in aquaculture. Algal biomass serves as a sustainable protein alternative, reducing pressure on wild fish stocks.
3. **Sustainability and Environmental Stewardship**: Algal farming can be conducted in a manner that aligns with environmental conservation goals. Algae cultivation sequesters carbon, mitigates pollution, and requires fewer natural resources than conventional feed crops, making it a highly sustainable option.
4. **Health and Welfare of Farmed Species**: Beyond its basic nutritional value, algae are rich in bioactive compounds that promote gut health, boost immunity, and enhance disease resistance in farmed species, reducing the need for antibiotics and other chemical interventions.

Exploring the Types of Algae for Aquafeeds

Algae can be broadly classified into two main categories—microalgae and macroalgae (seaweeds)—both of which have demonstrated significant potential for use in aquafeeds.

- **Microalgae**: These are microscopic, single-celled organisms that thrive in a wide range of aquatic environments. Species such as *Spirulina*, *Chlorella*, *Nannochloropsis*, and *Tetraselmis* are particularly rich in proteins, essential fatty acids, pigments (e.g., carotenoids), and antioxidants. Microalgae are frequently included in aquafeeds for their superior nutritional value and their role in promoting growth, health, and immunity in fish and shrimp.
- **Macroalgae (Seaweeds)**: Macroalgae are larger, multicellular organisms commonly found in marine environments. Seaweed species such as *Ulva*, *Gracilaria*, *Laminaria*, and *Sargassum* are rich in carbohydrates, vitamins, minerals, and functional compounds like polysaccharides (e.g., alginates, carrageenans, and agar). These seaweeds are often incorporated into aquafeeds to improve digestion, boost immune function, and enhance the overall welfare of farmed species.

The Increasing Demand for Algal Biomass in Aquaculture

As the aquaculture industry faces mounting pressure to adopt more sustainable practices, the demand for alternative feed ingredients such as algal biomass is expected to grow. Algal-based feeds are being increasingly recognized not only for their sustainability but also for their ability to contribute to fish growth performance, survival rates, and overall health. Furthermore, the use of algal biomass aligns with the industry's goals to reduce its reliance on marine resources, lower its carbon footprint, and promote eco-friendly farming practices.

This chapter provides a comprehensive exploration of the potential of algal biomass as a sustainable feed ingredient in aquaculture. It will cover the types of algae that are most suitable for aquafeeds, their nutritional profiles, the specific health benefits they offer to farmed species, their environmental advantages over traditional feed ingredients, and the challenges and opportunities for their large-scale adoption in the industry.

2. Types of Algal Biomass

Algal biomass is broadly classified into two main categories: microalgae and macroalgae (seaweeds). Each type of algae brings its own distinct nutritional benefits and applications in aquafeeds, making them valuable for enhancing the diets of farmed aquatic species.

Microalgae

Microalgae are single-celled organisms known for their high nutrient content and ease of cultivation. Commonly used species in aquaculture include *Chlorella*, *Spirulina*, and *Nannochloropsis*. These microalgae are rich in proteins, essential fatty acids (like eicosapentaenoic acid [EPA] and docosahexaenoic acid [DHA]), vitamins, and pigments such as carotenoids and chlorophyll.

Microalgae offer several key benefits in aquafeeds:

- **High protein content:** Some microalgae contain protein levels comparable to or higher than conventional sources like soy or fishmeal. This is particularly beneficial in meeting the dietary protein requirements of carnivorous fish species.
- **Essential fatty acids:** EPA and DHA, which are vital for the health and growth of many fish species, are naturally present in microalgae. These long-chain omega-3 fatty acids are often limited in plant-based diets but are essential for promoting healthy immune systems and optimal growth rates in aquatic animals.

- **Pigments and antioxidants:** Microalgae are a source of carotenoids, such as astaxanthin and β-carotene, which not only improve the coloration of fish and shrimp but also function as antioxidants to protect cells from oxidative damage.

Microalgae are commonly used in the early life stages of fish, as larvae and juvenile fish benefit significantly from their high-quality nutrients. Additionally, microalgae-based feeds help enhance the nutritional value of fish for human consumption, making them an important component in both fish health and food quality.

Macroalgae (Seaweeds)

Macroalgae, or seaweeds, are multicellular, larger algae species that grow in marine and freshwater environments. Some of the most commonly cultivated macroalgae species for aquafeeds include *Ulva* (sea lettuce), *Laminaria* (kelp), and *Gracilaria*. These seaweeds are typically rich in carbohydrates, polysaccharides, minerals, and bioactive compounds, offering complementary nutritional benefits to microalgae.

Macroalgae provide the following benefits in aquafeeds:

- **Carbohydrate and fiber content:** Seaweeds contain significant amounts of polysaccharides such as alginate, agar, and carrageenan, which serve as prebiotics that support gut health in fish and shrimp. These fibers can promote beneficial gut microbiota and enhance immune responses.
- **Bioactive compounds:** Many seaweeds are known to contain bioactive compounds that offer health benefits to aquatic species, such as antimicrobial, anti-inflammatory, and antioxidant properties. These compounds can help reduce the incidence of disease and improve the overall health of farmed fish.
- **Minerals and trace elements:** Seaweeds are naturally rich in minerals like iodine, potassium, calcium, and magnesium, which are important for bone development, metabolic functions, and overall growth in aquatic species.

3. Nutritional Composition of Algal Biomass

Algal biomass is gaining widespread recognition as a highly nutritious and sustainable feed ingredient for aquaculture due to its diverse and rich nutrient profile. It provides a balanced combination of essential proteins, lipids, vitamins, minerals, and bioactive compounds, making it a valuable alternative to traditional feed ingredients like fishmeal and plant-based proteins. The exceptional nutritional content of algal biomass supports the health, growth, and productivity of aquatic species, and contributes to their overall well-being.

Proteins

Algal biomass is particularly valued for its high protein content, which can range from 20% to 70% of dry weight, depending on the species and growing conditions. Microalgae such as *Spirulina* and *Chlorella* are among the most protein-rich, containing up to 70% crude protein. This makes them comparable to, and in some cases superior to, conventional protein sources like fishmeal and soybean meal.

Key aspects of the protein content in algal biomass:

- **Essential amino acids:** Algal proteins are well-balanced in essential amino acids, including lysine, methionine, and arginine, which are often limiting in plant-based feed ingredients. This makes algae an excellent complement to other feed ingredients, improving the overall amino acid profile of the diet.
- **Digestibility:** Many studies have shown that the proteins in algae are highly digestible for fish, particularly when processed and incorporated into aquafeeds. Digestibility can be enhanced further through fermentation, enzymatic treatment, or other pre-processing methods, ensuring efficient nutrient utilization by the fish.

This high-quality protein makes algal biomass a promising substitute for fishmeal, particularly in carnivorous fish species that have high protein requirements. Algae-based feeds can support optimal growth, tissue repair, and muscle development in aquaculture systems.

Lipids and Fatty Acids

Algae are a rich source of essential lipids and fatty acids, particularly polyunsaturated fatty acids (PUFAs), which are crucial for maintaining fish health and supporting various physiological functions. Some microalgae, such as *Nannochloropsis* and *Schizochytrium*, are renowned for their high content of long-chain omega-3 fatty acids, including eicosapentaenoic acid (EPA) and docosahexaenoic acid (DHA).

- **Omega-3 and Omega-6 fatty acids:** These essential fatty acids are important for the development of fish tissues, including brain and nervous system functions. They also play a vital role in modulating immune responses and reducing inflammation in fish, thereby improving their overall resilience to diseases.
- **Lipid content:** The lipid content of algal biomass can range from 5% to 50%, depending on the species and cultivation conditions. Algal oils, particularly those rich in omega-3 fatty acids, are highly valued as they can replace fish oil, another finite resource, in aquafeeds.

Fish fed algal-based diets rich in omega-3 PUFAs have been shown to exhibit improved growth performance, better feed conversion ratios, and enhanced immune function. Additionally, the inclusion of algae in aquafeeds can enhance the nutritional value of fish for human consumption, increasing the levels of beneficial omega-3 fatty acids in fish fillets.

Vitamins and Minerals

Algal biomass is an excellent source of essential vitamins and minerals that are vital for fish growth, development, and overall health. These micronutrients play important roles in metabolic processes, immune function, and disease prevention.

- **Vitamins:** Algae are particularly rich in fat-soluble vitamins, including vitamins A, D, E, and K, which support various physiological functions in fish. Vitamin E, for example, acts as an antioxidant, protecting cells from oxidative damage, while vitamin D supports bone development and calcium regulation. B-complex vitamins, including B12, thiamine, and riboflavin, are also abundant in algae and are crucial for energy metabolism and red blood cell formation.
- **Minerals:** Algae contain significant amounts of essential minerals such as iodine, selenium, zinc, and iron. Iodine, abundant in seaweeds, is important for thyroid function and metabolism, while selenium and zinc play key roles in immune responses and antioxidant defense systems. These trace elements are often deficient in conventional plant-based feed ingredients, making algae a valuable addition to aquafeeds for supplying these crucial nutrients.

Bioactive Compounds

Beyond their nutrient content, algae are rich in bioactive compounds that provide additional health benefits to aquatic species. These bioactive components include antioxidants, antimicrobial peptides, and polysaccharides, which help to improve fish health, boost immune function, and protect against diseases.

- **Antioxidants:** Algae contain several potent antioxidants, such as phycocyanin (in *Spirulina*) and carotenoids like astaxanthin and β-carotene. These compounds help neutralize free radicals and reduce oxidative stress in fish, which can occur due to factors like environmental stress, disease, or poor water quality.
- **Chitin and polysaccharides:** Seaweeds, in particular, are a rich source of polysaccharides such as alginate, carrageenan, and laminarin. These compounds have prebiotic effects, promoting the growth of beneficial gut bacteria and improving digestive health. Chitin, found in the cell

walls of certain algae, also supports immune function and enhances disease resistance.

These bioactive compounds contribute to the overall health and well-being of farmed fish, improving their resilience to stress and disease while supporting optimal growth and feed efficiency. Incorporating algal biomass into aquafeeds can thus provide both nutritional and functional benefits, improving the sustainability and productivity of aquaculture systems.

Algal biomass offers a comprehensive and balanced nutritional profile, providing high-quality proteins, essential fatty acids, vitamins, minerals, and bioactive compounds. These nutrients support the health, growth, and productivity of farmed fish and shrimp, making algae a highly promising and sustainable feed ingredient for the aquaculture industry. The combination of its nutritional value and functional properties highlights the potential of algal biomass to improve the overall performance and sustainability of aquaculture operations, while also addressing global challenges related to feed resource scarcity and environmental impact.

4. Health Benefits of Algal Biomass in Aquaculture

Algal biomass is increasingly recognized not only for its nutritional value but also for its range of health-promoting properties in aquaculture. Incorporating algae into fish and shrimp diets goes beyond meeting basic nutritional needs; it delivers functional benefits that can enhance growth, immune function, gut health, and overall disease resistance.

4.1 Improved Growth Performance

Algal biomass, particularly from species like *Spirulina* and *Chlorella*, provides high-quality proteins, essential amino acids, and fatty acids, which are key to improving growth performance in farmed fish and shrimp. These nutrients boost feed conversion ratios (FCR), allowing fish to efficiently convert feed into body mass. Studies have demonstrated that fish fed algal-supplemented diets exhibit faster growth rates and improved weight gain compared to those on conventional feeds. The inclusion of essential fatty acids, such as omega-3 (DHA and EPA), further promotes healthy tissue development, growth, and optimal reproductive performance, making algal biomass a valuable tool for enhancing aquaculture productivity.

4.2 Enhanced Immune Function

Algae are rich in bioactive compounds like carotenoids (e.g., astaxanthin and beta-carotene), polysaccharides, and vitamins that play crucial roles in supporting the immune system of aquatic species. These bioactive compounds can stimulate both the innate and adaptive immune responses, helping fish and shrimp better fend off infections. For instance, polysaccharides in algae act as immunostimulants, enhancing the activity of macrophages, lymphocytes, and phagocytic cells that are involved in pathogen defense. Fish and shrimp raised on algal-based diets have shown improved resistance to common aquaculture diseases, including bacterial infections caused by *Aeromonas* and *Vibrio* species, as well as viral and parasitic diseases. By strengthening the immune system, algae help reduce the reliance on antibiotics and chemical treatments, fostering a healthier and more sustainable aquaculture environment.

4.3 Gut Health

The health of the gastrointestinal system plays a pivotal role in the overall well-being and growth of aquatic species. Certain types of algae, such as *Ulva* and *Chlorella*, contain polysaccharides and fiber that exhibit prebiotic properties. These compounds encourage the growth of beneficial gut microbiota, which aids in nutrient digestion and absorption, leading to better feed efficiency and nutrient utilization. A balanced gut microbiota also contributes to enhanced immunity and reduced susceptibility to pathogens. The presence of prebiotics in algae promotes gut barrier function, making it more difficult for harmful bacteria to colonize the gut lining. This improvement in gut health can lead to better growth performance and overall resilience of farmed species, ultimately translating to higher aquaculture productivity.

4.4 Antioxidant Protection

Algal biomass is abundant in antioxidants, such as carotenoids, phycocyanin, and polyphenols, which play a critical role in protecting fish and shrimp cells from oxidative stress. Oxidative stress, often caused by environmental factors such as poor water quality, high stocking densities, and temperature fluctuations, can lead to cellular damage and a weakened immune system. Antioxidants from algae neutralize reactive oxygen species (ROS) and free radicals, preventing cellular damage and improving the resilience of farmed species under stressful conditions. This antioxidant protection contributes to better survival rates, healthier animals, and improved stress tolerance, particularly in intensive aquaculture systems where environmental challenges are more common.

The inclusion of algal biomass in aquaculture diets offers numerous health benefits that extend beyond basic nutrition. By promoting growth, enhancing immune function, supporting gut health, and providing antioxidant protection, algae contribute to the overall welfare and productivity of farmed aquatic species. As the aquaculture industry continues to shift toward more sustainable and functional feed ingredients, algal biomass is poised to play a pivotal role in the future of aquafeeds.

5. Environmental Benefits of Algal Biomass in Aquafeeds

One of the most compelling reasons for the increasing interest in algal biomass as a feed ingredient in aquaculture is its potential to provide significant environmental benefits. As the global demand for sustainable feed solutions rises, algae offer a range of ecological advantages that make them a low-impact, resource-efficient alternative to traditional feed sources like fishmeal and soy.

5.1 Low Resource Input

Algae stand out as a highly sustainable feed ingredient because they can be cultivated with minimal resource inputs. Unlike conventional crops used in aquafeeds, such as soybeans and grains, algae do not require arable land or large amounts of freshwater. They can be grown in non-arable, marginal lands, including deserts or coastal areas, and in saline or brackish water, making them an ideal solution for regions where freshwater resources are scarce or agricultural land is limited. This characteristic significantly reduces the competition for resources between food production for human consumption and feed production for livestock and aquaculture. In a world where water scarcity and land degradation are growing concerns, algae offer a resource-efficient alternative that aligns with the principles of sustainable agriculture and aquaculture.

5.2 Carbon Sequestration and Climate Change Mitigation

Algae are photosynthetic organisms that play an essential role in carbon sequestration. During their growth, algae absorb large amounts of carbon dioxide (CO□) from the atmosphere and convert it into biomass through photosynthesis. This process not only reduces the overall carbon footprint of algal farming but also contributes to mitigating climate change by capturing atmospheric CO□. In large-scale cultivation systems, algae can act as natural carbon sinks, offsetting greenhouse gas emissions from other industries. Additionally, the use of algae in aquafeeds can indirectly reduce the carbon footprint of aquaculture by decreasing reliance on fishmeal and soy, which are associated with higher levels of CO□ emissions due to overfishing and land-use changes (e.g., deforestation for soybean production).

5.3 Reduced Pressure on Fishmeal and Soy

One of the most pressing challenges facing the aquaculture industry is its reliance on fishmeal and soy, which are the primary protein sources in conventional aquafeeds. Fishmeal, made from wild-caught fish, is increasingly unsustainable due to overfishing and the depletion of marine ecosystems. Similarly, soybean cultivation has been linked to deforestation, habitat destruction, and biodiversity loss, particularly in regions like the Amazon rainforest. The incorporation of algal biomass into aquafeeds provides a sustainable alternative that can help reduce the industry's dependence on these environmentally damaging feed ingredients. Algae offer a protein source that is nutritionally comparable to fishmeal and soy, making them a viable replacement in many aquafeed formulations. By reducing the demand for fishmeal, algae can alleviate pressure on wild fish stocks and contribute to the preservation of marine biodiversity. Likewise, replacing soy with algal biomass can help combat deforestation and its associated environmental impacts.

5.4 Waste Utilization and Nutrient Recycling

Algae have the unique ability to grow on a wide variety of nutrient sources, including wastewater from aquaculture and agriculture. This capability makes them an excellent candidate for recycling nutrients and reducing environmental pollution. In integrated aquaculture systems, algae can be cultivated using nutrient-rich effluents from fish farms, which often contain high levels of nitrogen and phosphorus. These nutrients, if left untreated, can contribute to water pollution and eutrophication in surrounding ecosystems. By utilizing waste streams as a nutrient source, algae can help close the nutrient loop in aquaculture, converting potentially harmful waste into valuable biomass for feed production. This circular approach not only reduces the environmental impact of aquaculture but also enhances the overall sustainability of the production system.

5.5 Algae's Role in a Circular Economy

The cultivation of algae within a circular economy framework is gaining momentum as industries look for ways to minimize waste and maximize resource efficiency. Algal farming can be integrated into various industrial processes, where it serves as both a nutrient sink and a producer of valuable biomass. For example, algae can be grown on agricultural runoff, capturing excess nutrients and preventing them from entering water bodies. The harvested algae can then be processed into feed ingredients, biofuels, or other bioproducts, creating a closed-loop system that promotes sustainability. This integration of algae into circular economy models aligns with global efforts to reduce waste, conserve resources, and create more resilient food systems. In aquaculture, the use of algae as a feed ingredient supports these goals by offering a renewable, low-impact alternative to traditional feed sources.

The environmental benefits of algal biomass in aquafeeds are vast and far-reaching. Algae's low resource requirements, ability to sequester carbon, and potential to reduce reliance on unsustainable feed ingredients like fishmeal and soy make them an ideal candidate for sustainable aquaculture. Additionally, algae's capacity to utilize waste and recycle nutrients contributes to a more circular, resource-efficient system, minimizing the environmental impact of aquaculture practices. As the aquaculture industry continues to expand, embracing innovative feed solutions like algal biomass will be crucial for reducing its ecological footprint and ensuring long-term sustainability.

6. Challenges and Limitations

While algal biomass offers numerous advantages as a sustainable, nutrient-rich feed ingredient, its widespread adoption in aquaculture is hindered by several key challenges. Understanding these limitations is essential for advancing research and innovation aimed at overcoming these obstacles.

6.1 Cost of Production

One of the most significant challenges facing the large-scale use of algal biomass in aquafeeds is the high cost of production. Cultivating algae, whether microalgae or macroalgae, requires controlled environmental conditions, specific growth media, and often complex harvesting and processing techniques. These factors contribute to the relatively high production costs compared to conventional feed ingredients like fishmeal, soy, and other plant-based materials.

Several factors contribute to the cost issue:

- **Energy Requirements:** Algal cultivation systems, especially closed systems like photobioreactors, often require substantial energy inputs for lighting, mixing, and temperature control. Open-pond systems, while more cost-effective, are prone to contamination and may not produce the same high-quality biomass.
- **Nutrient Inputs:** Algae require specific nutrients for growth, such as nitrogen and phosphorus. Providing these nutrients at scale can be costly, especially if high-quality, consistent biomass is required.
- **Harvesting and Processing:** Efficient harvesting of microalgae can be technically challenging, as they are typically suspended in large volumes of water. Additionally, drying, extracting, or otherwise processing the biomass to make it suitable for use in aquafeeds adds to production expenses.

However, efforts are being made to lower the cost of algae production through advancements in cultivation technologies, such as photobioreactors, which can enhance growth rates while using fewer resources. Additionally, genetic engineering of algal strains to optimize nutrient utilization and increase productivity is a promising area of research. As these technologies mature, the cost of producing algal biomass is expected to decrease, making it a more economically viable option for large-scale aquafeed production.

6.2 Nutrient Variability

Another challenge in using algal biomass in aquafeeds is the variability in its nutritional composition. The nutrient profile of algae, including protein, lipid, and fatty acid content, can vary significantly depending on the species, environmental conditions, and cultivation methods. Factors such as light intensity, nutrient availability, temperature, and salinity all influence the growth and nutrient content of algae.

For example:

- **Microalgae** such as *Chlorella* and *Spirulina* are known for their high protein content, but this can fluctuate based on how they are grown.
- **Macroalgae** (seaweeds), such as *Ulva* or *Laminaria*, tend to be rich in carbohydrates and polysaccharides, but their nutrient concentrations can vary widely across different batches.

This inconsistency poses a challenge for aquafeed manufacturers who require standardized, predictable feed formulations to ensure optimal fish growth and health. Fluctuations in nutrient content could lead to inconsistent performance in farmed species, making it difficult to rely solely on algal biomass without careful formulation adjustments. To mitigate this issue, continued research into optimizing growth conditions and developing specific algal strains with stable and enhanced nutrient profiles is necessary. Controlled cultivation environments, such as closed bioreactors, and genetically modified algae could help create more consistent and reliable algal-based feed products.

6.3 Processing and Digestibility

A major limitation of some algae species in aquafeeds is the presence of indigestible or anti-nutritional compounds that can reduce the digestibility of the biomass when consumed by fish or shrimp. Certain algae contain polysaccharides, such as cellulose or other complex carbohydrates that are difficult for many aquatic species to break down.

For instance

- **Microalgae** can have cell walls composed of tough, indigestible components like chitin and cellulose, which can reduce nutrient bioavailability and limit the overall feed efficiency.
- **Macroalgae**, especially seaweeds, often contain high levels of polysaccharides (e.g., alginate, carrageenan) that are not easily digested by fish.

To address these digestibility issues, various processing techniques are being explored, such as:

- **Fermentation**: This technique uses microbial action to break down indigestible compounds in the algae, enhancing their digestibility. Fermentation can also enrich algal biomass with beneficial probiotics and other bioactive compounds.
- **Enzymatic treatments**: These involve using enzymes to degrade the tough cell walls of microalgae, increasing the bioavailability of proteins, lipids, and other essential nutrients.
- **Mechanical or chemical processing**: This includes grinding, drying, or extracting key nutrients from algae to create more digestible feed components.

These processing techniques aim to improve nutrient utilization and make algae a more attractive feed ingredient, though they add additional steps and costs to the production process.

6.4 Scale and Supply Chain

The global supply chain for algal biomass suitable for aquafeeds is still in its early stages of development. While the market for microalgae in human nutrition, cosmetics, and pharmaceuticals is relatively well-established, the scale required for widespread use in aquaculture is much larger and more complex.

- **Scale of Production**: Current algal production facilities are often geared towards smaller-scale, high-value markets (e.g., health supplements), rather than the bulk feed markets required by the aquaculture industry.
- **Supply Chain Development**: Developing an efficient and reliable supply chain for algal biomass, particularly one capable of producing large quantities at competitive prices, remains a challenge. Ensuring year-round availability and maintaining consistent quality are critical factors for its adoption in aquaculture.

Collaborative efforts between research institutions, governments, and industry stakeholders are essential to build the infrastructure needed to scale up algal biomass production for aquafeeds.

Despite the challenges, algal biomass holds immense potential as a sustainable, nutrient-rich alternative for use in aquafeeds. Overcoming cost barriers, improving nutrient consistency, addressing digestibility issues, and scaling up production are critical steps toward the widespread adoption of algae in the aquaculture industry. As research advances and new technologies emerge, it is likely that algal biomass will become a more feasible and attractive option, contributing to the long-term sustainability and resilience of global aquaculture.

7. Future Prospects

The future of algal biomass in aquafeeds is bright, driven by increasing awareness of its nutritional benefits and environmental sustainability. As research and technology advance, several promising developments are poised to enhance the role of algal biomass in aquaculture.

7.1 Genetic Engineering and Breeding

Advancements in genetic engineering and selective breeding are expected to revolutionize algal biomass production by creating strains with improved characteristics that meet the specific needs of aquaculture. Key areas of focus include:

- **Enhanced Nutrient Profiles**: Through genetic modification, researchers can develop algal strains that are richer in essential nutrients, such as proteins, omega-3 fatty acids, and vitamins. This would ensure that algal biomass can more effectively meet the nutritional requirements of various fish and shrimp species.
- **Faster Growth Rates**: Genetic engineering techniques can also accelerate the growth rates of algae, allowing for higher yields within shorter cultivation periods. This increase in productivity is crucial for making algal biomass a competitive feed ingredient.
- **Improved Stress Tolerance**: Breeding and genetic modification can help create algal strains that are more resilient to fluctuating environmental conditions (e.g., temperature, salinity). Such adaptability can enhance biomass production in diverse aquaculture settings, including those affected by climate change.

These advancements are likely to lead to more consistent and reliable algal products for aquaculture, ultimately supporting better fish health and growth performance.

7.2 Integration with Circular Economy

The concept of the circular economy is gaining traction in sustainable aquaculture practices, and algal biomass fits seamlessly into this model. Potential avenues for integration include:

- **Utilizing Aquaculture Waste**: Algal cultivation can be combined with aquaculture systems to utilize waste materials, such as nutrient-rich effluents from fish farms, as growth media for algae. This approach not only reduces the environmental impact of waste disposal but also provides a sustainable nutrient source for algal growth.
- **Nutrient Recycling**: By leveraging waste streams from agriculture and aquaculture, the circular economy can enhance nutrient recycling. Algae can absorb excess nutrients, such as nitrogen and phosphorus, thereby mitigating the risk of eutrophication in water bodies and contributing to healthier ecosystems.
- **Co-Product Development**: Integrating algal biomass into a broader circular economy framework allows for the development of various co-products, such as biofuels, fertilizers, and bioplastics, thereby diversifying revenue streams and reducing waste in aquaculture systems.

This synergistic relationship not only enhances sustainability but also supports economic viability within the aquaculture sector.

7.3 Nutritional Synergies

The future of algal biomass in aquafeeds also lies in its potential for nutritional synergies when combined with other sustainable feed ingredients. For example:

- **Insect-Based Protein**: Combining algal biomass with insect protein can create a highly nutritious and balanced feed formulation. Insects are rich in protein and amino acids, while algae provide essential fatty acids and bioactive compounds. Together, they can fulfill the nutritional requirements of fish and shrimp, potentially enhancing growth rates and health outcomes.
- **Plant-Based Ingredients**: Algae can be used in conjunction with other plant-based feed ingredients, such as legumes and grains, to optimize nutrient profiles while minimizing reliance on conventional sources like fishmeal and soy. This combination can provide a more comprehensive amino acid profile and enhance overall feed quality.

Exploring these synergies will contribute to the development of more sustainable and efficient aquafeeds, further reducing the ecological footprint of aquaculture.

Table: Overview of Algal Biomass as a Superfood for Aquaculture

Aspect	Description
Types of Algal Biomass	- **Microalgae**: Single-celled organisms (e.g., Spirulina, Chlorella) rich in proteins, essential fatty acids, and pigments.
	- **Macroalgae (Seaweeds)**: Multicellular organisms (e.g., Ulva, Laminaria) high in carbohydrates, polysaccharides, and bioactive compounds.
Nutritional Composition	- **Proteins**: Up to 70% crude protein; comparable to fishmeal; provides all essential amino acids.
	- **Lipids**: Rich in omega-3 and omega-6 fatty acids (e.g., EPA, DHA), crucial for growth and immune function.
	- **Vitamins and Minerals**: Source of vitamins A, C, D, E, and B-complex, and trace minerals like iodine, selenium, and zinc.
	- **Bioactive Compounds**: Contains antioxidants (e.g., phycocyanin, carotenoids) that reduce oxidative stress and enhance immune function.
Health Benefits	- **Improved Growth Performance**: Enhances feed conversion efficiency and weight gain.
	- **Enhanced Immune Function**: Boosts immune responses and resistance to infections, reducing the need for antibiotics.
	- **Gut Health**: Prebiotic properties improve beneficial gut microbiota, aiding nutrient digestion and absorption.
	- **Antioxidant Protection**: Rich in antioxidants, protecting cells from oxidative damage and enhancing resilience against environmental stressors.
Environmental Benefits	- **Low Resource Input**: Grows in non-arable land and saline water, minimizing competition for freshwater and agricultural land.
	- **Carbon Sequestration**: Absorbs CO2 during photosynthesis, contributing to climate change mitigation.
	- **Reduced Pressure on Fishmeal and Soy**: Decreases reliance on fishmeal and soy, which have significant environmental footprints.
	- **Waste Utilization**: Can be cultivated using wastewater, recycling nutrients and reducing pollution.
Challenges and Limitations	- **Cost of Production**: Higher costs compared to conventional feed ingredients, though technology advancements may reduce costs over time.
	- **Nutrient Variability**: Nutrient composition varies by species and cultivation conditions, creating challenges for consistent aquafeed formulations.
	- **Processing and Digestibility**: Some algae contain indigestible compounds; processing techniques are needed to improve digestibility.
Future Prospects	- **Genetic Engineering and Breeding**: Development of strains with improved nutrient profiles, growth rates, and stress tolerance.

	- **Integration with Circular Economy**: Utilizing waste from aquaculture to grow algae, promoting sustainability.
	- **Nutritional Synergies**: Combining algal biomass with other sustainable feed ingredients (e.g., insect protein) for enhanced nutritional benefits.

8. Conclusion

Algal biomass holds significant promise as a "superfood" for aquaculture, offering a sustainable and nutrient-rich alternative to traditional feed ingredients. Its high protein content, essential fatty acids, bioactive compounds, and environmental benefits make it an ideal candidate for modern aquafeeds. While challenges related to cost and variability remain, ongoing research and technological innovations are expected to drive the broader adoption of algal biomass in aquaculture systems, contributing to a more sustainable, resilient, and productive industry. As the aquaculture industry continues to expand in response to global seafood demand, integrating algal biomass into feed formulations will play a crucial role in meeting this demand while safeguarding the planet's resources. By leveraging the unique advantages of algal biomass, the industry can align itself with sustainable practices that enhance fish health and productivity while minimizing environmental impact. The future of aquaculture is undoubtedly intertwined with the use of innovative and sustainable feed solutions like algal biomass. As researchers, industry stakeholders, and policymakers collaborate to address existing challenges and harness the full potential of algae, the promise of a more sustainable aquaculture industry will become a reality, ensuring food security for future generations.

References

Bhatnagar, A., & Devi, P. (2013). Wastewater treatment using algae: A review. Critical Reviews in Environmental Science and Technology, 43(2), 127-168.

Boussiba, S. (2000). Carotenogenesis in the green alga Dunaliella bardawil: Effects of light and salinity. Journal of Applied Phycology, 12(3), 287-297.

Campbell, P. J., & Prichard, J. R. (2018). An overview of algal biomass in aquaculture: Nutritional profiles and applications. Aquaculture Nutrition, 24(1), 4-17.

Cordero, M. A., & Palacios, A. J. (2020). Microalgae as a source of nutrients for aquaculture. Journal of Applied Phycology, 32(3), 1611-1626.

Del Campo, J. A., Rodriguez, H., & Vargas, M. A. (2007). Microalgae as a source of polyunsaturated fatty acids in aquaculture. Aquaculture Research, 38(9), 947-955.

El-Baky, H. A., & El-Baroty, G. S. (2010). Nutritional evaluation of some green microalgae as a source of protein in aquaculture. Journal of Applied Phycology, 22(6), 701-708.

FAO (2020). The State of World Fisheries and Aquaculture 2020. Food and Agriculture Organization of the United Nations.

Finkel, Z. V., & Irwin, A. J. (2008). Nutrient limitation of phytoplankton growth in the ocean. Nature, 452(7188), 16.

Gao, K., & McKinley, K. (2019). Seaweed biofuels: A sustainable energy source for the future. Nature Sustainability, 2(6), 559-563.

Gouveia, L., & Oliveira, A. C. (2009). Microalgae as a source of biofuels and other value-added products. Food Technology and Biotechnology, 47(2), 135-144.

Hu, Q., & Guo, J. (2020). Microalgae for aquaculture feed: Nutritional and environmental aspects. Aquaculture Reports, 17, 100383.

Kalscheur, K. F., & Minton, M. (2021). Feeding the world with algal biofuels: The aquaculture connection. Sustainable Energy & Fuels, 5(3), 505-515.

Koller, M., & Rumpold, B. A. (2013). Microalgae: A new food for the future. Food Research International, 54(1), 109-123.

Lembke, J., & Wirth, S. (2020). Nutritional value of microalgae in aquaculture feed: A systematic review. Aquaculture Nutrition, 26(2), 259-270.

Makkar, H. P. S., & Becker, K. (2010). Nutritional value of microalgae for animal feed. Animal Feed Science and Technology, 156(1-2), 1-15.

Manns, M., & Lang, B. (2018). The potential of microalgae in aquaculture feed. Marine Drugs, 16(2), 38.

Naylor, R. L., & Burke, M. (2005). Aquaculture and ocean resources: Raising tigers of the sea. Proceedings of the National Academy of Sciences, 102(6), 1992-1997.

Parvez, M. H., & Nishi, N. (2017). Utilization of microalgae in aquaculture: Current status and future prospects. Fish Physiology and Biochemistry, 43(5), 1271-1282.

Renaud, S. M., & Thinh, L. V. (2015). Microalgae in aquaculture: A review of their nutritional and health benefits. Aquaculture Nutrition, 21(4), 617-634.

Rhim, J. W., & Lee, J. S. (2016). Utilization of algal biomass in aquaculture: Health benefits and nutritional value. Journal of Food Science, 81(6), 1239-1247.

Rosenberg, J. N., & Leitz, J. (2016). Microalgae as a sustainable feed ingredient for aquaculture: A review. Aquaculture Reports, 4, 48-57.

Rumpold, B. A., & Schlüter, O. K. (2013). Nutritional and sensory properties of microalgae in food products. Comprehensive Reviews in Food Science and Food Safety, 12(2), 175-192.

Sarker, P. K., & Becker, K. (2018). Nutritional value of microalgae and their potential for aquaculture feed: A review. Aquaculture Research, 49(2), 355-373.

Zhang, Y., & Li, Y. (2017). Microalgae in aquaculture: A new perspective on nutrition and health. Aquaculture Research, 48(2), 593-608.

Zhou, Q., & Chen, Y. (2021). Biomass production and application of microalgae in aquaculture. Bioresource Technology Reports, 15, 100663.

9

Nutrient Bioavailability in Aquafeeds Enhancing Efficiency

1. Introduction

Aquaculture has emerged as one of the fastest-growing food production sectors worldwide, a trend driven by several factors, including the rising global demand for seafood, the depletion of wild fish stocks, and the search for sustainable protein sources. As the human population continues to grow and dietary preferences shift towards healthier, protein-rich food options, aquaculture plays a pivotal role in addressing food security. However, the expansion of aquaculture also brings forth significant challenges, particularly in the efficient use of resources, including feed. One of the primary challenges for the aquaculture industry is ensuring that fish and other aquatic organisms receive optimal nutrition through formulated feeds. Aquafeeds are typically designed to provide a balanced mix of proteins, lipids, carbohydrates, vitamins, and minerals necessary for the growth, health, and reproduction of farmed species. However, not all nutrients in these feeds are equally digestible or absorbable by the animals. The efficiency with which nutrients are utilized depends largely on their bioavailability—a term that refers to the proportion of ingested nutrients that are available for absorption and utilization by the body.

1.1 Defining Nutrient Bioavailability in Aquafeeds

In aquaculture nutrition, nutrient bioavailability is a crucial determinant of overall feed efficiency. It is defined as the fraction of a nutrient that, after ingestion, becomes available for utilization in physiological processes such as growth, tissue repair, metabolism, and reproduction. In practice, this means that bioavailable nutrients must first be digested in the gastrointestinal tract, absorbed through the intestinal walls, transported to various tissues, and finally, metabolized or incorporated into the body. Nutrient bioavailability directly affects the performance of farmed aquatic species. For example, if a significant portion of a nutrient is poorly digested or absorbed, it is excreted rather than utilized, leading to inefficient growth and increased feed costs. Additionally, undigested nutrients released into the environment contribute to water pollution, causing eutrophication and other ecological imbalances. Therefore, improving nutrient bioavailability is not only essential for optimizing production efficiency but also for mitigating the environmental footprint of aquaculture.

1.2 The Importance of Bioavailability for Growth and Sustainability

Optimizing nutrient bioavailability in aquafeeds is fundamental to achieving the dual goals of sustainable aquaculture and cost-effective production. For aquaculture producers, maximizing the conversion of feed into growth (often measured as the feed conversion ratio, FCR) is vital for profitability. A low FCR means that fish require less feed to gain body mass, reducing the overall cost of feed—a major expense in aquaculture operations. Conversely, if bioavailability is low, a higher amount of feed is required, which not only raises production costs but also leads to greater environmental concerns due to the accumulation of uneaten feed and waste. Furthermore, nutrient bioavailability has implications for the health and well-being of aquatic animals. Essential nutrients such as amino acids, fatty acids, vitamins, and minerals must be efficiently absorbed to support vital physiological processes, including immune function, stress tolerance, and reproductive success. Poor bioavailability of key nutrients can compromise fish health, leading to increased susceptibility to diseases, slower growth, and higher mortality rates. In light of these challenges, research and innovation are focused on improving bioavailability through various approaches, including the selection of high-quality feed ingredients, advanced feed processing techniques, and the use of functional additives. Such strategies aim to enhance the digestibility and absorption of nutrients, allowing fish to derive maximum nutritional benefit from their diets.

1.3 Overview of the Chapter

This chapter delves into the various aspects of nutrient bioavailability in aquafeeds, providing an in-depth understanding of the factors influencing bioavailability and the methods used to enhance nutrient absorption in aquaculture systems. The following sections will cover:

- **Factors Affecting Nutrient Bioavailability:** A discussion of the biological, dietary, and environmental factors that impact how well nutrients are absorbed and utilized by aquatic species. This includes species-specific digestive capacities, the role of feed composition, and interactions between nutrients.
- **Strategies for Enhancing Nutrient Bioavailability:** This section will explore cutting-edge approaches to improve the bioavailability of nutrients, including advancements in feed processing (e.g., extrusion,

fermentation), the use of feed additives (e.g., enzymes, probiotics, chelated minerals), and ingredient selection (e.g., alternative protein sources like plant and insect-based meals).

- **Benefits of Improved Bioavailability:** A look at the benefits of enhancing nutrient bioavailability, including better growth performance, improved fish health, reduced environmental impact, and cost savings for producers.
- **Challenges and Future Prospects:** The chapter concludes by addressing the ongoing challenges in improving bioavailability, including cost considerations, variability in ingredient quality, and the need for species-specific solutions. It will also highlight future prospects, such as the integration of advanced technologies and precision nutrition in the development of next-generation aquafeeds.

2. Understanding Nutrient Bioavailability

Nutrient bioavailability refers to the proportion of nutrients in feed that are digested, absorbed, and metabolized by the body to support growth, health, and other physiological processes. In aquaculture, ensuring high nutrient bioavailability is crucial for optimizing feed efficiency, reducing costs, and minimizing environmental impact. Bioavailability encompasses several stages, including digestion in the gastrointestinal tract, absorption through the intestinal lining, transportation to target tissues, and subsequent metabolism. Fish, as aquatic animals, have unique physiological and environmental constraints that influence nutrient bioavailability. Factors such as species, age, digestive capabilities, feeding habits, water temperature, and feed composition all play a role in determining how well nutrients are absorbed and utilized.

Key Nutrients in Aquafeeds

Aquafeeds are formulated to deliver a balance of essential macronutrients (proteins, lipids, carbohydrates) and micronutrients (vitamins and minerals) necessary for the growth and health of fish and other aquatic species. The bioavailability of these nutrients determines how effectively fish can convert the feed into biomass and energy.

1. **Proteins:** Proteins are the most critical macronutrient in aquafeeds, serving as the building blocks for muscle tissue, enzymes, hormones, and other vital molecules. Efficient protein bioavailability is essential for optimal growth and feed conversion. Proteins in aquafeeds come from various sources, including fishmeal, plant proteins, and alternative sources like insect meal. The digestibility of these protein sources varies significantly, with fishmeal often having the highest bioavailability due

to its amino acid profile that closely matches the requirements of fish. However, plant-based proteins may have lower bioavailability due to anti-nutritional factors (ANFs) such as phytates and tannins.

2. **Lipids:** Lipids provide both energy and essential fatty acids such as omega-3 (EPA, DHA) and omega-6, which are crucial for cell membrane integrity, immune function, and reproductive health. The bioavailability of lipids is influenced by the type of fatty acids present and how they are incorporated into the feed. For example, fish oil has a high bioavailability due to its high content of long-chain polyunsaturated fatty acids (LC-PUFAs), which are easily metabolized by fish. Plant-based lipids, while beneficial, may not offer the same bioavailability unless supplemented with additional sources of EPA and DHA.
3. **Carbohydrates:** While fish are not as efficient as land animals in utilizing carbohydrates, they still serve as a supplementary energy source. Carbohydrates help spare protein for growth by providing energy for basic metabolic functions. However, the bioavailability of carbohydrates in fish depends on the species, as carnivorous fish tend to have lower carbohydrate digestion capabilities compared to herbivorous or omnivorous species. Proper feed formulation ensures that carbohydrates are provided in a digestible form, such as starch, to optimize their utilization.
4. **Vitamins and Minerals:** Vitamins (e.g., A, D, E, B-complex) and minerals (e.g., calcium, phosphorus, selenium) play vital roles in metabolic processes, bone formation, and immune responses. The bioavailability of these micronutrients is influenced by their chemical form, interactions with other nutrients, and the processing methods used in feed production. For instance, chelated minerals (minerals bound to organic molecules) tend to have higher bioavailability compared to their inorganic counterparts, as they are more easily absorbed in the intestines.

Factors Affecting Nutrient Bioavailability

Several factors impact how efficiently nutrients are absorbed and utilized in fish diets. Understanding and addressing these factors is essential for improving feed formulations and enhancing the bioavailability of key nutrients.

1. **Feed Composition**: The quality and source of feed ingredients significantly affect nutrient bioavailability. For example, animal-derived proteins like fishmeal and blood meal generally have higher bioavailability compared to plant-based proteins, which may contain ANFs that inhibit nutrient absorption. Ingredients like soybean meal and corn gluten meal are commonly used in aquafeeds but may require

treatments such as heat processing or enzyme supplementation to reduce ANFs and improve nutrient digestibility.

2. **Feed Processing**: The processing techniques used in aquafeed production, such as extrusion, pelleting, and drying, influence the physical and chemical properties of the feed, which in turn affects nutrient availability. For instance, extrusion, a common processing method, increases the digestibility of starch and proteins by breaking down their complex structures. However, excessive heat treatment during processing may lead to nutrient losses, particularly heat-sensitive vitamins like vitamin C.
3. **Species-Specific Factors**: Different fish species have varying digestive capacities, influencing their ability to utilize specific nutrients. Carnivorous fish, such as salmon and trout, have evolved to digest and absorb proteins and lipids more efficiently than carbohydrates. In contrast, omnivorous and herbivorous species like tilapia and carp can utilize a wider range of nutrients, including plant-based proteins and carbohydrates. Therefore, feed formulations must be tailored to the specific nutrient requirements and digestive capabilities of the target species.
4. **Enzyme Activity**: The presence and activity of digestive enzymes are critical for breaking down complex nutrients into simpler forms that can be absorbed. Endogenous enzymes produced by the fish play a central role in digesting proteins, fats, and carbohydrates. In addition, exogenous enzymes, such as phytase and protease, are often added to aquafeeds to improve the breakdown of anti-nutritional compounds and enhance nutrient absorption. For example, phytase helps release bound phosphorus in plant-based feeds, making it more bioavailable for fish.

Nutrient bioavailability is a complex but essential aspect of aquaculture nutrition, determining how efficiently fish and other aquatic species can utilize the nutrients provided in their feeds. Understanding the factors that affect nutrient absorption, such as feed composition, processing methods, and species-specific digestive capacities is key to formulating high-performance aquafeeds. By enhancing nutrient bioavailability, aquaculture producers can improve growth rates, reduce feed costs, and minimize environmental impacts, paving the way for a more sustainable and productive industry.

3. Strategies to Enhance Nutrient Bioavailability

Enhancing nutrient bioavailability in aquafeeds is crucial for maximizing feed efficiency and supporting optimal growth, health, and sustainability in aquaculture. By improving nutrient absorption, aquaculture operations can reduce feed wastage, lower costs, and decrease environmental impact. Several approaches are available to improve nutrient bioavailability, each targeting specific aspects of feed formulation, processing, and nutrient utilization.

3.1. Feed Ingredient Selection and Quality

The choice of feed ingredients directly affects the bioavailability of nutrients. High-quality ingredients with balanced amino acids, essential fatty acids, and bioavailable micronutrients can promote efficient nutrient uptake. The quality of the feed ingredients and their composition must be carefully considered to avoid anti-nutritional factors (ANFs) that can hinder nutrient absorption.

- **Fishmeal**: Fishmeal has long been considered the gold standard in aquafeeds due to its high digestibility and complete amino acid profile. It provides superior bioavailability of essential nutrients such as proteins, lipids, and micronutrients. However, concerns over the sustainability of fishmeal have driven efforts to explore alternative protein sources. Substitutes like plant-based meals, insect proteins, and microalgae are increasingly being used, though they require careful formulation to ensure nutrient bioavailability is comparable to fishmeal.
- **Plant-based Proteins**: While plant-based proteins such as soybean meal, corn gluten meal, and pea protein are sustainable alternatives to fishmeal, they often contain ANFs like phytates, tannins, and protease inhibitors. These compounds can bind essential minerals, inhibit protein digestion, and reduce nutrient absorption. Techniques such as fermentation, enzymatic treatment, and selective breeding of low-ANF crops are being developed to improve the digestibility and bioavailability of plant-based proteins.
- **Insect Proteins and Algae**: Insect-based proteins and algal biomass offer promising alternatives to conventional protein sources in aquafeeds. These ingredients are highly digestible and can be tailored to specific nutritional needs. However, the consistency of nutrient profiles and cost of production are challenges that must be addressed to fully capitalize on their potential.

3.2 Feed Processing Techniques

Feed processing methods have a significant impact on nutrient bioavailability by altering the structure and composition of feed. Proper processing can improve the digestibility of complex nutrients, while over-processing can lead to nutrient degradation.

- **Extrusion and Pelleting**: These methods involve applying heat, moisture, and pressure to shape the feed and improve its texture. Extrusion, in particular, enhances the digestibility of proteins and starches by breaking down their molecular structures. However, excessive heat during processing can cause the loss of heat-sensitive nutrients like vitamins, amino acids, and fatty acids. Therefore, it is essential to optimize processing conditions to retain nutrient integrity while improving digestibility.
- **Microencapsulation**: This innovative technique involves coating nutrients with protective layers to prevent their degradation during feed processing, storage, and digestion. Microencapsulation is particularly useful for protecting essential fatty acids, vitamins, probiotics, and other bioactive compounds that are sensitive to environmental stressors. This method enhances nutrient stability and ensures that nutrients remain bioavailable during digestion.

3.3. Use of Digestive Enzymes

Digestive enzymes play a crucial role in breaking down complex feed components, enhancing nutrient digestion, and improving bioavailability. Exogenous enzymes are often added to aquafeeds to complement the endogenous digestive enzymes produced by fish.

- **Phytase**: Phytate is an ANF present in many plant-based feed ingredients that binds phosphorus, making it unavailable for absorption. Phytase is an enzyme that breaks down phytate, releasing bioavailable phosphorus and improving the utilization of other essential minerals like calcium, zinc, and magnesium. By adding phytase to plant-based diets, aquaculture producers can reduce the need for inorganic phosphorus supplements, lowering both costs and environmental phosphorus discharge.
- **Proteases**: Protease enzymes break down complex protein structures into simpler peptides and amino acids, improving protein digestion and utilization. This is especially beneficial in plant-based diets, where protein digestibility is often lower than in animal-based feeds. Supplementing proteases can enhance the efficiency of protein use, leading to improved growth performance and feed conversion ratios in farmed fish.

3.4. Functional Feed Additives

In addition to enzymes, several functional feed additives can enhance nutrient bioavailability by supporting digestion, absorption, and nutrient utilization in aquatic animals.

- **Probiotics and Prebiotics**: Probiotics are live beneficial microorganisms that improve gut health by promoting a healthy microbiota. A balanced gut microbiota enhances nutrient digestion and absorption, particularly in the intestine where most nutrients are absorbed. Prebiotics, such as inulin and fructooligosaccharides, are non-digestible food ingredients that stimulate the growth of beneficial bacteria in the gut. Together, probiotics and prebiotics support a healthy digestive system, leading to improved nutrient bioavailability.
- **Chelated Minerals**: Chelated minerals are bound to organic molecules such as amino acids, peptides, or polysaccharides, which enhances their absorption in the gastrointestinal tract. Compared to inorganic mineral salts, chelated minerals are more bioavailable and less likely to interact with other dietary components that could inhibit their absorption. Chelated forms of zinc, copper, iron, and other essential trace minerals are commonly used in aquafeeds to enhance growth, immune function, and overall health in fish.
- **Essential Oils**: Derived from plants, essential oils contain bioactive compounds such as phenols, terpenes, and aldehydes, which have been shown to improve digestive enzyme activity, gut health, and immune responses. When added to aquafeeds, essential oils can enhance nutrient digestion and absorption by promoting a more efficient gastrointestinal environment. Some essential oils also possess antimicrobial properties, reducing the likelihood of infections that could impair nutrient uptake.

Enhancing nutrient bioavailability in aquafeeds is a multifaceted challenge that involves selecting high-quality feed ingredients, optimizing feed processing techniques, and incorporating functional additives such as enzymes, probiotics, and chelated minerals. By improving nutrient absorption and utilization, aquaculture operations can achieve higher growth rates, better feed conversion efficiency, and reduced environmental impact. The use of innovative strategies such as microencapsulation, fermentation, and enzyme supplementation is critical to overcoming the limitations associated with plant-based proteins and other alternative feed ingredients. Ultimately, improving nutrient bioavailability will contribute to a more sustainable, cost-effective, and resilient aquaculture industry.

4. Benefits of Enhancing Nutrient Bioavailability

Enhancing nutrient bioavailability in aquafeeds is a pivotal strategy that brings numerous advantages to the aquaculture industry. By optimizing the efficiency with which fish and other aquatic species utilize nutrients, aquaculture producers can not only improve the productivity of their operations but also contribute to sustainability and better animal health. The following sections outline the key benefits associated with enhancing nutrient bioavailability.

4.1 Improved Growth Performance

One of the most significant benefits of enhancing nutrient bioavailability is the improvement in growth performance of farmed fish. When nutrients are more readily absorbed and utilized by the body, fish grow faster and more efficiently. This translates to better feed conversion ratios (FCR)—the amount of feed required to produce a unit of body weight in fish. A lower FCR indicates that fish are converting feed into body mass more effectively, resulting in:

- **Faster growth rates**: Enhanced nutrient absorption ensures that fish receive sufficient amounts of proteins, lipids, and essential micronutrients to support rapid muscle development and overall growth.
- **Optimized production cycles**: Faster-growing fish can reach market size more quickly, allowing producers to shorten production cycles and increase the number of harvests per year.
- **Improved yield**: With better nutrient utilization, fish grow uniformly and achieve optimal size, which enhances the yield and quality of the final product.

4.2 Reduced Feed Costs

Feed costs represent a major expense in aquaculture, often accounting for over 50% of total production costs. Improving nutrient bioavailability allows producers to reduce the overall amount of feed required to meet the nutritional needs of fish. This results in significant cost savings for aquaculture operations by:

- **Minimizing feed wastage**: When nutrients are better absorbed, less feed is wasted in the form of uneaten or undigested materials. This leads to more efficient feed use and lower feed costs.
- **Maximizing nutrient utilization**: Enhancing bioavailability ensures that fish derive the maximum benefit from the nutrients present in their feed, reducing the need for higher feeding rates.
- **Reducing dependence on expensive ingredients**: By improving the digestibility of alternative protein sources, such as plant-based or insect-based proteins, producers can rely less on costly fishmeal and fish oil, contributing to overall cost reduction.

4.3 Better Fish Health

The health and well-being of farmed fish are closely tied to the quality and bioavailability of the nutrients they receive. Enhancing nutrient bioavailability provides fish with the essential vitamins, minerals, and fatty acids they need for optimal immune function, disease resistance, and overall vitality. The benefits for fish health include:

- **Strengthened immune response**: Bioavailable nutrients such as zinc, vitamin C, and omega-3 fatty acids play critical roles in supporting immune system function. With enhanced nutrient uptake, fish are better equipped to resist infections and recover from diseases.
- **Reduced dependence on antibiotics**: By improving overall health and disease resistance, aquaculture producers can reduce the need for antibiotics and other medical treatments. This not only benefits fish welfare but also helps prevent the development of antibiotic-resistant bacteria, contributing to a more sustainable aquaculture industry.
- **Enhanced stress tolerance**: Proper nutrient bioavailability ensures that fish have the necessary resources to cope with environmental stressors such as temperature fluctuations, poor water quality, and overcrowding, which can otherwise compromise their health and growth.

4.4 Environmental Sustainability

Sustainability is a critical concern for the global aquaculture industry, which faces increasing scrutiny over its environmental impact. Improving nutrient bioavailability can contribute to more sustainable practices by minimizing the environmental footprint of aquaculture operations. Key environmental benefits include:

- **Reduced feed wastage and pollution**: When nutrients are absorbed more efficiently, there is less uneaten or undigested feed in the water. This reduces the release of organic waste, including nitrogen and phosphorus, into the surrounding environment. Excessive nutrient discharge from aquaculture systems can lead to water pollution and eutrophication, which degrades water quality and harms aquatic ecosystems.
- **Lower nutrient loading in water bodies**: By enhancing bioavailability, aquafeeds can be formulated to supply the precise amount of nutrients required by fish, minimizing the risk of nutrient leaching into water bodies. This helps maintain a balanced aquatic ecosystem and reduces the likelihood of harmful algal blooms and oxygen depletion in the water.
- **Support for sustainable feed sources**: Enhancing the digestibility and nutrient absorption of alternative feed ingredients, such as plant-based proteins or insect meals, reduces the reliance on wild-caught fish for fishmeal and fish oil. This supports the sustainable management of marine resources and helps conserve biodiversity.

4.5. Economic and Industry Benefits

Beyond the direct operational advantages, improving nutrient bioavailability in aquafeeds can positively impact the broader aquaculture industry and contribute to economic growth. Some of the overarching benefits include:

- **Higher profitability**: With improved growth performance, lower feed costs, and reduced disease outbreaks, aquaculture producers can increase their profitability and long-term viability.
- **Increased production efficiency**: Enhanced nutrient bioavailability allows for more efficient production, meaning that aquaculture systems can produce more fish using fewer resources. This contributes to greater food security by meeting the growing demand for seafood with fewer inputs.
- **Innovation and technology integration**: The focus on improving nutrient bioavailability has led to the development of innovative feed additives, processing techniques, and feed formulation technologies. These advancements contribute to the overall modernization of the aquaculture industry, fostering a more resilient and competitive sector.

Enhancing nutrient bioavailability in aquafeeds is a powerful tool for improving growth performance, reducing feed costs, promoting fish health, and advancing environmental sustainability in aquaculture. By adopting strategies that optimize nutrient digestion and absorption, aquaculture producers can increase the efficiency of their operations while minimizing their environmental footprint. As the global demand for seafood continues to rise, the ability to improve nutrient bioavailability will play a crucial role in shaping the future of sustainable aquaculture.

5. Challenges and Limitations

Despite the promising advancements in enhancing nutrient bioavailability in aquafeeds, several challenges and limitations need to be addressed for broader adoption and practical implementation. Understanding these constraints is crucial for developing effective solutions that benefit the entire aquaculture sector.

5.1. Cost of Functional Additives

One of the primary challenges in enhancing nutrient bioavailability is the cost associated with incorporating functional additives such as enzymes, probiotics, prebiotics, and chelated minerals. While these additives can significantly improve nutrient utilization and fish health, they also add to the overall production costs of the feed. This can be a barrier for small- and medium-scale aquaculture producers who operate on tighter profit margins.

- **High enzyme costs**: Enzymes like phytase and protease are effective at improving nutrient digestion but can be expensive, especially when used at high inclusion levels.
- **Investment in advanced processing technologies**: Techniques like microencapsulation, precision grinding, or extrusion require specialized equipment and expertise, which may not be feasible for all producers.
- **Increased feed prices**: The added cost of functional additives can lead to higher feed prices, making the feeds less accessible to producers in low-income regions, thereby limiting the widespread adoption of bioavailability-enhancing strategies.

5.2 Inconsistent Results Across Species and Diets

The effectiveness of bioavailability-enhancing strategies can vary widely depending on fish species, dietary composition, and environmental conditions. While some strategies, such as enzyme supplementation, may yield excellent results in one species, they may be less effective or even counterproductive in others.

- **Species-specific nutrient requirements**: Different fish species have varying digestive capabilities and nutrient absorption efficiencies. For example, carnivorous fish, such as salmon and trout, have different dietary protein and lipid needs compared to herbivorous or omnivorous species like tilapia and carp.
- **Variability in diet composition**: Diets with high levels of anti-nutritional factors (ANFs), such as those based on plant proteins, require specific additives to mitigate their effects. The presence of different ANFs, such as tannins, saponins, and lectins, can influence nutrient digestibility in unpredictable ways.
- **Environmental factors**: Water quality, temperature, and pH can also impact nutrient absorption and enzyme activity, leading to inconsistent results in real-world aquaculture settings.

5.3 Nutrient Interactions

Nutrient interactions are a complex factor that can either enhance or reduce the bioavailability of specific nutrients in aquafeeds. For example, the presence of certain minerals or anti-nutritional factors can negatively interact with essential nutrients, reducing their bioavailability.

- **Mineral-mineral interactions**: Excess calcium in the diet can interfere with the absorption of phosphorus, which is a critical nutrient for skeletal development and energy metabolism in fish.

- **Anti-nutritional factors (ANFs)**: Compounds like phytates, tannins, and oxalates, which are commonly found in plant-based ingredients, can bind to proteins and minerals, making them unavailable for absorption.
- **Lipid and vitamin interactions**: Fat-soluble vitamins (A, D, E, and K) require lipids for absorption. An imbalance in lipid levels or poor lipid digestibility can reduce the bioavailability of these essential vitamins.

5.4 Stability of Additives during Feed Processing

Another limitation is the stability of functional additives during feed processing and storage. Heat-sensitive nutrients and additives, such as certain vitamins, fatty acids, and probiotics, can degrade during processing methods like extrusion and pelleting, reducing their effectiveness.

- **Thermal degradation**: High temperatures during feed processing can lead to the loss of heat-sensitive nutrients, reducing their bioavailability and nutritional value.
- **Oxidation of lipids**: Essential fatty acids, particularly omega-3 and omega-6, are prone to oxidation, which reduces their bioavailability and can generate harmful by-products.
- **Microencapsulation costs**: Although microencapsulation can protect sensitive nutrients, it adds to the cost and complexity of feed production.

6. Future Prospects

The future of enhancing nutrient bioavailability in aquafeeds looks promising, with ongoing research and innovation focusing on novel feed ingredients, advanced processing technologies, and functional additives. Several emerging trends and technologies are expected to shape the future of nutrient bioavailability in aquaculture.

6.1 Precision Nutrition

Advances in precision nutrition involve tailoring diets to the specific nutritional needs of different fish species, life stages, and production conditions. The use of machine learning, data analytics, and nutrient profiling can enable the development of highly customized feed formulations.

- **Dynamic feed formulation**: Real-time data on fish growth, water quality, and nutrient digestibility can be integrated to create diets that optimize bioavailability and minimize waste.
- **Species-specific enzyme blends**: Customized enzyme blends that target specific anti-nutritional factors or dietary components for particular fish species can improve nutrient uptake.

6.2 Exploration of Novel Protein Sources

Continued research into sustainable and digestible protein sources such as insect meals, algae, and single-cell proteins is expected to provide new opportunities for enhancing nutrient bioavailability in aquafeeds. These novel ingredients can be specifically processed and supplemented to maximize nutrient utilization.

- **Insect-based proteins**: With high digestibility and bioactive compounds, insect meals can improve nutrient absorption and gut health.
- **Algal proteins**: Algae offer a rich source of highly digestible proteins and essential fatty acids that can enhance nutrient bioavailability and reduce reliance on fishmeal.

6.3 Development of Customized Enzyme Blends

Enzyme technology is expected to evolve further, with the development of personalized enzyme blends that target specific dietary components or fish species. These blends will optimize the breakdown of complex nutrients and mitigate the effects of anti-nutritional factors.

- **Multi-enzyme formulations**: Combining proteases, carbohydrases, and phytases in tailored formulations can improve the digestibility of plant-based proteins and enhance overall feed efficiency.
- **Bioinformatics-driven enzyme design**: Advances in bioinformatics and genomics can aid in designing enzymes with higher specificity and stability under varying pH and temperature conditions.

Table: Nutrient Bioavailability in Aquafeeds: Enhancing Efficiency

Category	Details
Definition	The proportion of nutrients in feed that is available for absorption and utilization by the fish's body.
Key Nutrients in Aquafeeds	**Proteins**: Essential for growth, tissue repair, and enzyme synthesis. **Lipids**: Provide energy and essential fatty acids (e.g., omega-3, omega-6). **Carbohydrates**: Serve as an energy source but are less efficiently utilized by fish compared to land animals. **Vitamins & Minerals**: Support metabolism, bone formation, and immune responses.
Factors Affecting Nutrient Bioavailability	- **Feed composition**: Source and quality of ingredients; presence of anti-nutritional factors (e.g., phytates, tannins). - **Feed processing**: Methods like extrusion and pelleting impact digestibility. - **Species-specific factors**: Different fish species have varying abilities to digest and absorb nutrients. - **Enzyme activity**: Endogenous and exogenous enzymes improve nutrient breakdown.

Strategies to Enhance Bioavailability	- **Feed Ingredient Selection**: Use of high-quality ingredients (e.g., fishmeal, plant-based proteins with processing to reduce anti-nutritional factors). - **Feed Processing Techniques**: Extrusion, pelleting, microencapsulation to improve digestibility. - **Exogenous Enzymes**: Use of enzymes like phytase and protease to enhance nutrient digestion. - **Functional Feed Additives**: Probiotics, prebiotics, chelated minerals, and essential oils improve digestion and nutrient absorption.
Benefits of Enhancing Bioavailability	- **Improved growth performance**: Faster growth rates, better Feed Conversion Ratio (FCR). - **Reduced feed costs**: Less feed required for growth, cost savings. - **Better fish health**: Enhanced immune function, reduced need for antibiotics. - **Environmental sustainability**: Reduced waste output, lower nutrient pollution.
Challenges & Limitations	- **Cost of functional additives**: Enzymes, probiotics, and chelated minerals can increase production costs. - **Inconsistent results**: Variability in outcomes based on species, diet, and environment. - **Nutrient interactions**: Certain nutrients may inhibit absorption of others.
Future Prospects	- **Precision nutrition**: Use of advanced formulation technologies and machine learning to develop species-specific diets. - **Alternative proteins**: Research on insect meals, algae, and single-cell proteins for digestible, bioavailable nutrients. - **Personalized enzyme blends**: Development of tailored enzymes for optimizing digestion of specific nutrients in different species.

7. Conclusion

Enhancing nutrient bioavailability in aquafeeds is essential for optimizing the growth, health, and sustainability of farmed aquatic species. By focusing on feed ingredient selection, advanced processing techniques, and the use of functional additives such as enzymes and probiotics, the aquaculture industry can significantly improve nutrient absorption and utilization. This not only leads to better growth performance and feed efficiency but also reduces environmental impacts, supports animal health, and enhances the economic viability of aquaculture systems. Future innovations in precision nutrition, novel protein sources, and enzyme technologies will further enhance nutrient bioavailability, contributing to the sustainable expansion of the aquaculture industry.

References

Adeoye, A. A., Jaramillo-Torres, A., & Davies, S. J. (2018). Effect of dietary enzyme supplementation on growth performance, nutrient utilization and gut histology of Atlantic salmon (Salmo salar) fed diets with graded levels of soy protein concentrate. Aquaculture Nutrition, 24(1), 226-235.

Adeoye, A. A., Jaramillo-Torres, A., Fox, S. W., Merrifield, D. L., & Davies, S. J. (2016). Supplementation of exogenous enzymes and probiotic on enzyme activities, intestinal histo-morphology and microbiome of European sea bass (Dicentrarchus labrax) fed plant-based diets. Aquaculture Nutrition, 22(1), 101-110.

Barrows, F. T., Gaylord, T. G., Sealey, W. M., Porter, L., & Smith, C. E. (2008). Supplementation of plant-based diets for rainbow trout (Oncorhynchus mykiss) with macro and trace minerals. Aquaculture Nutrition, 14(5), 466-474.

Barrows, F. T., Stone, D. A., & Hardy, R. W. (2007). The effects of extrusion conditions on the nutritional value of soybean meal for rainbow trout (Oncorhynchus mykiss). Aquaculture, 265(1-4), 244-252.

Bureau, D. P., & Hua, K. (2010). Towards effective nutritional management of waste outputs in aquaculture, with particular reference to salmonid aquaculture operations. Aquaculture Research, 41(5), 777-792.

Cheng, Z. J., & Hardy, R. W. (2004). Nutritional requirements of tilapia: practical implications for aquafeeds. In: C. Lim & C. Webster (Eds.), Tilapia: Biology, Culture, and Nutrition (pp. 387-418). Food Products Press.

Cho, C. Y., & Bureau, D. P. (1997). Reduction of waste output from salmonid aquaculture through feeds and feeding. Progressive Fish-Culturist, 59(2), 155-160.

Cowey, C. B., & Luquet, P. (1983). Physiological basis of protein requirements of fish. Critical Reviews in Aquatic Sciences, 1(2), 163-187.

Drew, M. D., Borgeson, T. L., & Thiessen, D. L. (2007). A review of processing of feed ingredients to enhance diet digestibility in finfish. Animal Feed Science and Technology, 138(2), 118-136.

Francis, G., Makkar, H. P. S., & Becker, K. (2001). Antinutritional factors present in plant-derived alternate fish feed ingredients and their effects in fish. Aquaculture, 199(3-4), 197-227.

Gatlin, D. M., & Li, P. (2004). Dietary additives in fish feeds: current and future applications. Aquaculture Magazine, 30(2), 51-55.

Gatlin, D. M., & Li, P. (2008). Additives in aquafeeds: Are they needed and effective? Aquaculture Magazine, 34(2), 57-63.

Gatlin, D. M., & Phillips, R. (2007). Dietary supplementation of exogenous enzymes for fish culture. Aquaculture Nutrition, 13(1), 77-84.

Gaylord, T. G., & Barrows, F. T. (2009). Apparent digestibility of gross nutrients from feedstuffs in extruded diets for rainbow trout (Oncorhynchus mykiss). Aquaculture Nutrition, 15(2), 112-120.

Glencross, B. D. (2009). Reduced water oxygen availability and its effects on dietary digestible protein and energy utilisation efficiency in barramundi (Lates calcarifer). Aquaculture Research, 40(7), 750-759.

Glencross, B. D., Booth, M., & Allan, G. L. (2007). A feed is only as good as its ingredients – a review of ingredient evaluation strategies for aquaculture feeds. Aquaculture Nutrition, 13(1), 23-34.

Glencross, B. D., Hawkins, W. E., & Curnow, J. G. (2003). Nutritional and commercial limitations of ingredients for farmed fish feeds. Aquaculture Nutrition, 9(4), 175-185.

Hardy, R. W. (2010). Utilization of plant proteins in fish diets: effects of global demand and supplies of fishmeal. Aquaculture Research, 41(5), 770-776.

Hardy, R. W., & Barrows, F. T. (2002). Diet formulation and manufacture. In: J. E. Halver & R. W. Hardy (Eds.), Fish Nutrition (pp. 505-600). Academic Press.

Hardy, R. W., & Barrows, F. T. (2002). Fishmeal-free feeds for salmon and trout. Aquaculture Magazine, 28(3), 46-49.

Hossain, M. A., Nahar, N., & Kamal, M. (1997). Nutrient digestibility coefficients and digestive enzyme activities in rohu (Labeo rohita) juveniles fed diets containing various plant and animal proteins. Aquaculture Research, 28(1), 43-49.

Karapanagiotidis, I. T., Psofakis, P., & Mente, E. (2008). Digestibility and bioavailability of amino acids in fishmeal and alternative protein sources for fish. Reviews in Fisheries Science, 16(2), 140-156.

Kaushik, S. J., & Seiliez, I. (2010). Protein and amino acid nutrition and metabolism in fish: current knowledge and future needs. Aquaculture Research, 41(3), 322-332.

Krogdahl, Å., Bakke-McKellep, A. M., & Baeverfjord, G. (2003). Effects of graded levels of standard soybean meal on intestinal structure, mucosal enzyme activities, and pancreatic response in Atlantic salmon (Salmo salar L.). Aquaculture Nutrition, 9(6), 361-371.

Krogdahl, Å., Gajardo, K., & Kortner, T. M. (2015). Possible mechanisms of plant ingredient anti-nutrients in fish aquaculture, with emphasis on soy anti-nutrients. Aquaculture, 454, 136-150.

Krogdahl, A., Penn, M., & Bakke, A. M. (2015). Alternative protein sources and digestive function alterations in teleost fish. Proceedings of the Nutrition Society, 74(4), 317-328.

Kumar, V., & Makkar, H. P. S. (2016). Phytate: anti-nutritional factor in oilseed protein sources. In M. Gaber, H. Al-Harbi & H. El-Nakhlawy (Eds.), Recent Advances in Fish Nutrition (pp. 165-190). Cambridge Scholars Publishing.

Kumar, V., Makkar, H. P. S., & Becker, K. (2012). Detoxified Jatropha curcas kernel meal as a dietary protein source for common carp (Cyprinus carpio L.): growth performance, nutrient utilization and body composition. Aquaculture, 356-357, 105-110.

NRC (National Research Council). (1993). Nutrient Requirements of Fish. National Academies Press.

NRC (National Research Council). (2011). Nutrient Requirements of Fish and Shrimp. National Academies Press.

Olsen, R. E., Myklebust, R., Ringø, E., & Mayhew, T. M. (2001). The influence of dietary linseed oil and saturated fatty acids on caecal enterocytes in Arctic charr (Salvelinus alpinus L.): a quantitative ultrastructural study. Aquaculture, 194(1-2), 57-65.

Olsen, R. E., Ringø, E., & Henderson, R. J. (1998). The incorporation of lipids into the tissues of Arctic charr (Salvelinus alpinus L.) fed linseed oil, rapeseed oil or fish oil. Aquaculture Nutrition, 4(4), 243-252.

Silva, S. S. D. (1999). Nutritional research techniques for finfish: the South and Southeast Asian perspective. Aquaculture Nutrition, 5(1), 81-91.

Silva, S. S. D., & Anderson, T. A. (1995). Fish Nutrition in Aquaculture. Springer.

Vielma, J., & Ruohonen, K. (1999). Effects of dietary phosphorus on acid-base balance in juvenile rainbow trout (Oncorhynchus mykiss) in different water phosphorus concentrations. Aquaculture Nutrition, 5(1), 53-63.

10

Immunonutrition in Fish Feeding for Disease Resistance

1. Introduction

Aquaculture has rapidly emerged as a crucial industry to meet the growing global demand for seafood, contributing significantly to food security and economic development. However, as aquaculture operations expand and intensify, fish health management has become increasingly challenging. The close confinement of fish in high-density farming systems creates ideal conditions for the spread of infectious diseases. Stressors such as overcrowding, fluctuating water quality, and improper nutrition further exacerbate the susceptibility of fish to pathogens. As a result, disease outbreaks in aquaculture are more frequent and can lead to severe economic losses due to reduced productivity, increased mortality, and the cost of treatments. In the past, antibiotics and other pharmaceutical treatments were widely used to manage fish diseases. While these methods have been effective in controlling infections, they come with several significant drawbacks:

- **Antibiotic resistance**: Overuse of antibiotics has contributed to the rise of antibiotic-resistant bacteria, a major public health concern that can affect not only aquatic organisms but also humans through the food chain.
- **Residues in fish products**: Antibiotics and other chemicals may leave harmful residues in fish, posing potential health risks to consumers and limiting the export potential of farmed fish products to markets with stringent regulations.
- **Environmental contamination**: The use of chemicals in aquaculture can lead to the contamination of surrounding water bodies, affecting wild fish populations and the broader ecosystem.

Due to these limitations, there is a growing demand for sustainable and natural alternatives to traditional disease control methods. This has led to the increasing interest in immunonutrition—a concept that integrates nutrition with immune system enhancement in fish. Immunonutrition focuses on using specific nutrients and functional feed additives to boost the immune responses of fish, enabling them to resist infections and cope with environmental stressors without the need for excessive pharmaceutical interventions.

The Role of Immunonutrition in Aquaculture

Immunonutrition refers to the strategic use of nutrients, such as proteins, lipids, vitamins, minerals, and various bioactive compounds, to modulate the immune system of fish. By improving immune function, immunonutrition enhances the natural defense mechanisms of fish, helping them resist pathogenic threats and recover more quickly from infections. Nutrient-rich, well-formulated feeds not only support general growth and metabolic functions but also strengthen the immune response, enabling fish to better manage stress, diseases, and environmental challenges commonly encountered in aquaculture.

One of the key advantages of immunonutrition is that it represents a preventive approach to disease management. Instead of treating diseases after they occur, immunonutritional strategies aim to fortify the fish's immune system beforehand, reducing their susceptibility to infections and minimizing the impact of disease outbreaks. This preventive focus aligns with the broader movement toward sustainable aquaculture, which seeks to minimize the use of harmful chemicals and reduce the environmental footprint of fish farming.

Increasing Concern over Fish Health and Disease Management

The intensification of aquaculture, driven by the global population's demand for seafood, has led to several health management challenges:

1. **Disease outbreaks**: High stocking densities and rapid water flow in many aquaculture systems creates an ideal environment for the proliferation of bacteria, viruses, and parasites. Diseases such as viral hemorrhagic septicemia, infectious pancreatic necrosis, and vibriosis are among the most common pathogens affecting farmed fish.
2. **Stress**: Environmental stressors, including temperature fluctuations, poor water quality, and handling during routine farm operations, weaken fish immune systems, making them more susceptible to diseases.
3. **Nutritional Deficiencies**: Poorly formulated diets can contribute to deficiencies in essential nutrients required for immune function, such as vitamins, minerals, and specific amino acids, compromising fish health.

As disease management becomes more challenging, the traditional reliance on antibiotics and antimicrobials raises additional concerns:

- **Antibiotic resistance**: The overuse of antibiotics in aquaculture has led to the emergence of resistant bacterial strains, reducing the effectiveness of these drugs not only in treating fish diseases but also in human medicine.

- **Residue accumulation**: Antibiotics and chemical treatments can leave residues in fish tissues, posing health risks to consumers and creating obstacles for the export of farmed fish to international markets with strict safety standards.
- **Environmental impact**: Chemicals and antibiotics used in aquaculture may leach into surrounding aquatic ecosystems, negatively impacting wild fish populations and biodiversity.

As a result of these challenges, there is a growing need for alternative strategies that promote fish health and prevent diseases while maintaining environmental and food safety standards. Immunonutrition presents a promising solution by enhancing the natural disease resistance of fish through targeted nutritional interventions.

Immunonutrition: A Natural and Sustainable Solution

Immunonutrition provides a holistic and sustainable approach to disease prevention in aquaculture. By focusing on the dietary modulation of the immune system, immunonutrition not only supports the health and welfare of fish but also contributes to the sustainability of aquaculture by reducing the reliance on antibiotics and chemical treatments. Several key nutrients and bioactive compounds have been identified for their immunomodulatory effects in fish, including:

- Proteins and amino acids: Certain amino acids, such as arginine and glutamine, play vital roles in supporting immune cell function and tissue repair.
- Fatty acids: Omega-3 fatty acids, such as eicosapentaenoic acid (EPA) and docosahexaenoic acid (DHA), are known for their anti-inflammatory properties and their ability to modulate immune responses.
- Vitamins and minerals: Vitamins A, C, D, and E, as well as minerals like zinc and selenium, are essential for immune cell development, antioxidant protection, and the regulation of immune responses.
- Functional additives: Probiotics, prebiotics, plant-derived compounds (phytogenics), and other natural immunostimulants can enhance gut health, improve nutrient absorption, and support the fish's immune defense mechanisms.

The strategic inclusion of these nutrients in aquafeeds can strengthen the fish's ability to fight off infections, minimize disease-related losses, and enhance overall productivity. Furthermore, by improving immune function through nutrition, immunonutrition helps to reduce the need for antibiotics and other pharmaceuticals, thus addressing the challenges of antibiotic resistance and environmental contamination.

2. Fish Immune System: Overview

The fish immune system plays a critical role in protecting against pathogens and maintaining overall health. As aquatic organisms, fish are constantly exposed to waterborne pathogens, including bacteria, viruses, fungi, and parasites. To combat these threats, fish have evolved a complex immune system that consists of both innate and adaptive responses, similar to those found in mammals. However, due to differences in their physiology and aquatic environment, fish possess unique immune features that influence how they respond to infections and stressors.

Understanding the fundamental components of the fish immune system is essential for designing effective immunonutritional strategies. This section provides an overview of the key elements of the fish immune system, highlighting the interplay between innate immunity, which provides immediate protection, and adaptive immunity, which offers long-term, specific defense against pathogens.

2.1. Innate Immunity

Innate immunity serves as the first line of defense in fish, providing a rapid and non-specific response to a wide range of pathogens. It is an ancient and highly conserved immune mechanism that exists across all animal species, and in fish, it is especially important due to the aquatic environment, where pathogens can easily penetrate through the water.

The components of the innate immune system in fish can be classified into three main categories:

1. **Physical Barriers**: The physical barriers in fish, including the skin, scales, and mucus layers, are the first points of contact between the fish and the external environment. These barriers serve not only to prevent the entry of pathogens but also to trap and neutralize them.
    - **Skin and Scales**: The fish skin and scales provide a protective covering that shields internal tissues from mechanical damage and pathogen invasion. The skin is also home to immune cells that can detect and respond to potential threats.
    - **Mucus**: Fish secrete a thick mucus layer that covers their skin and gills. This mucus contains a variety of antimicrobial compounds,

such as lysozymes, lectins, and immunoglobulins, which actively neutralize pathogens. The mucus acts as a physical trap, preventing pathogens from attaching to the skin and gaining access to the internal organs.

2. **Cellular Defenses**: Cellular defenses form a critical part of the innate immune system. Various immune cells patrol the fish's body and respond to infections through processes like phagocytosis (engulfing pathogens) and cytotoxicity (destroying infected cells).
   - **Phagocytes**: Phagocytes, such as macrophages and neutrophils, are responsible for engulfing and digesting pathogens. These cells recognize pathogen-associated molecular patterns (PAMPs) on the surface of microorganisms and initiate an immediate immune response. Phagocytosis not only clears pathogens but also helps to present antigens to other immune cells, linking innate and adaptive immunity.
   - **Natural Killer (NK) Cells**: NK cells are involved in the destruction of virus-infected or cancerous cells. They play a role in recognizing abnormal cells and eliminating them before they can cause harm to the fish.
3. **Soluble Factors**: Soluble factors in fish blood and body fluids play an essential role in mediating the innate immune response. These factors circulate throughout the fish's body and help identify, neutralize, and eliminate pathogens.
   - **Complement System**: The complement system is a group of proteins that, once activated, form a cascade of reactions leading to the destruction of pathogens. The complement proteins can directly lyse microbial cells or tag them for recognition by phagocytes in a process known as opsonization.
   - **Lysozymes**: Lysozymes are enzymes present in fish mucus, skin, and other tissues. They target the bacterial cell wall, breaking it down and killing the bacteria. Lysozymes are a key component of the fish's innate immune defense against bacterial infections.
   - **Antimicrobial Peptides (AMPs)**: AMPs are small molecules that exhibit potent antimicrobial activity against a range of pathogens, including bacteria, viruses, and fungi. Fish produce a variety of AMPs that disrupt microbial membranes, leading to pathogen death.

The rapid and broad-spectrum nature of innate immunity is essential in the aquatic environment, where fish are continually exposed to potential pathogens.

Innate immune responses are not pathogen-specific, meaning that they provide immediate, generalized protection without the need for prior exposure to the pathogen.

2.2 Adaptive Immunity

While innate immunity provides an immediate response, adaptive immunity offers a more specific and long-lasting defense. Adaptive immunity in fish involves the recognition of specific antigens (foreign substances) and the development of a targeted response, which includes the production of antibodies and the activation of specialized immune cells. Although adaptive immunity in fish shares similarities with that of mammals, it is relatively less developed, particularly in lower vertebrate species.

Key components of adaptive immunity in fish include:

1. **B Cells and Antibody Production**: B cells are responsible for producing antibodies (immunoglobulins), which specifically bind to antigens on pathogens and neutralize them. Antibodies can also mark pathogens for destruction by other immune cells, a process known as opsonization. In fish, the primary immunoglobulin is IgM, which plays a central role in the humoral immune response. IgM circulates in the blood and other body fluids, binding to pathogens and facilitating their clearance from the body.
   - **Humoral Immunity**: The production of antibodies by B cells is the hallmark of humoral immunity. In fish, antibodies help combat extracellular pathogens (those outside of cells) such as bacteria and some parasites. However, the diversity and effectiveness of antibody responses in fish are generally lower than in mammals.
2. **T Cells and Cell-Mediated Immunity**: T cells are involved in cell-mediated immunity, which targets pathogens that have entered host cells, such as viruses and intracellular bacteria. Fish possess cytotoxic T cells, which can directly kill infected cells, as well as helper T cells, which coordinate the immune response by releasing cytokines that activate other immune cells.
   - **Cytotoxic T Cells**: These cells identify and destroy infected or abnormal cells by recognizing antigen fragments presented on the surface of these cells. This mechanism is crucial for controlling viral infections and some intracellular bacteria.
   - **Helper T Cells**: Helper T cells play a regulatory role in adaptive immunity by releasing cytokines that influence the activity of other immune cells, such as B cells and macrophages. These cells are

essential for orchestrating a balanced immune response and ensuring effective communication between innate and adaptive immunity.

3. **Immunological Memory**: One of the defining features of adaptive immunity is the development of immunological memory, which allows the immune system to remember previous encounters with pathogens. Memory B cells and T cells are generated following an infection or vaccination and enable the fish to mount a faster and more potent immune response upon re-exposure to the same pathogen.

While fish do possess memory cells, the adaptive immune system in fish is not as highly developed as in mammals, meaning that their ability to form long-lasting, robust immune memory is more limited. As a result, innate immunity plays a more prominent role in fish compared to mammals. This underlines the importance of supporting both the innate and adaptive immune systems through nutrition, particularly in farmed fish that are frequently exposed to stressful conditions and pathogenic threats.

The Importance of Innate Immunity in Fish

Due to the less advanced nature of adaptive immunity in fish, the innate immune system is often the primary line of defense, especially in early life stages when adaptive responses are still developing. This makes innate immune support through nutrition even more crucial. Fish rely heavily on their innate defenses to protect against pathogens, and dietary interventions that strengthen these defenses can greatly improve disease resistance and overall health.

By focusing on enhancing innate immunity, aquafeeds can improve fish resilience to infections, reduce the severity of disease outbreaks, and minimize the need for antibiotics or chemical treatments. This is where immunonutrition comes into play—targeting the nutritional needs of both the innate and adaptive immune systems to optimize health and performance in farmed fish. In the following sections, we will explore specific nutrients and feed additives that can enhance immune function in fish, contributing to better disease resistance, improved health outcomes, and sustainable aquaculture practices.

3. Nutrients That Influence Fish Immunity

In aquaculture, proper nutrition plays a vital role in boosting the immune system of fish, contributing to overall health and disease resistance. Certain nutrients, including proteins, lipids, vitamins, and minerals, are crucial for the normal functioning of the immune system. Additionally, specific feed additives such as probiotics, prebiotics, and plant-based compounds have been shown to enhance immune responses. This section explores the key nutrients that influence fish immunity, focusing on their immunomodulatory effects.

3.1 Proteins and Amino Acids

Proteins are the building blocks of cells, tissues, and immune molecules like antibodies and cytokines. In fish, adequate protein intake is necessary for immune cell function, and specific amino acids have been identified for their role in modulating immune responses. Some amino acids, such as arginine, glutamine, and taurine, act as immunonutrients by supporting various immune functions.

- **Arginine**: Arginine is a semi-essential amino acid that plays a pivotal role in enhancing the immune response. It is particularly known for its ability to stimulate the production of **nitric oxide (NO)**, a key signaling molecule in the immune system. Nitric oxide is involved in macrophage-mediated pathogen destruction, helping to neutralize bacteria, parasites, and viruses. Arginine also promotes T-cell proliferation and antibody production, contributing to a stronger immune response. In fish, dietary arginine supplementation has been shown to improve resistance to bacterial infections and enhance wound healing after tissue damage.
- **Glutamine**: Glutamine is a critical amino acid for the growth and function of immune cells, particularly during periods of stress or infection. It serves as an energy source for rapidly dividing cells, such as **enterocytes** in the intestines and **lymphocytes** in the immune system. By maintaining intestinal integrity, glutamine helps prevent pathogen translocation from the gut to the bloodstream, thus reducing the risk of systemic infections. Glutamine supplementation has been found to enhance fish immune responses, particularly under stress conditions like temperature fluctuations, hypoxia, or overcrowding.
- **Taurine**: Taurine is another immunomodulatory amino acid with **antioxidant** and **anti-inflammatory** properties. It helps regulate the production of cytokines, which are proteins that coordinate immune responses, and plays a key role in protecting cells from oxidative stress. In fish, taurine has been shown to improve survival rates during bacterial and viral infections, likely due to its role in reducing inflammation and protecting immune cells from damage.

3.2 Lipids and Fatty Acids

Lipids are not only a major energy source but also essential components of cell membranes, including those of immune cells. Fatty acids, particularly omega-3 and omega-6 fatty acids, play significant roles in modulating the immune response and inflammation.

- **Omega-3 Fatty Acids (EPA and DHA)**: The long-chain polyunsaturated fatty acids (PUFAs) **eicosapentaenoic acid (EPA)** and **docosahexaenoic acid (DHA)** are well known for their **anti-inflammatory** properties. Omega-3 fatty acids help reduce excessive inflammatory responses by decreasing the production of pro-inflammatory molecules such as prostaglandins and leukotrienes. This can prevent tissue damage during infections, leading to better health outcomes in fish. EPA and DHA are also important for maintaining the fluidity of cell membranes, which affects immune cell communication and the ability to recognize and respond to pathogens. Fish diets rich in omega-3 fatty acids have been associated with enhanced resistance to bacterial and parasitic infections.
- **Omega-6 Fatty Acids**: While omega-6 fatty acids are necessary for normal immune function, they promote **inflammatory responses** that help fight infections. However, an overabundance of omega-6 fatty acids relative to omega-3s can lead to chronic inflammation, which may increase the risk of tissue damage and immune dysregulation. Therefore, maintaining a balanced ratio of omega-3 to omega-6 fatty acids in aquafeeds is critical for optimizing immune homeostasis in fish.

3.3. Vitamins

Vitamins are essential micronutrients that participate in various immune processes, including antioxidant defense, immune cell activation, and pathogen recognition. Certain vitamins, such as vitamin A, vitamin C, vitamin D, and vitamin E, have been extensively studied for their immunomodulatory effects in fish.

- **Vitamin A**: Vitamin A plays a crucial role in maintaining **mucosal immunity** and the integrity of epithelial tissues, such as skin and gut linings, which are the first barriers against pathogen invasion. It also regulates the differentiation of immune cells, such as T cells and B cells, which are important for adaptive immunity. In fish, vitamin A deficiency can lead to impaired immune responses and increased susceptibility to infections. Adequate levels of vitamin A in aquafeeds are necessary to enhance resistance to pathogens and improve overall health.
- **Vitamin C**: Vitamin C acts as a powerful **antioxidant**, protecting immune cells from oxidative stress caused by reactive oxygen species (ROS) generated during infections. It also enhances the activity of phagocytic cells, such as macrophages and neutrophils, and stimulates antibody production. Vitamin C is particularly important for fish under stress or during periods of rapid growth, when immune function may be compromised. Supplementing fish diets with vitamin C has been shown

to improve resistance to a variety of pathogens, including bacteria and parasites.

- **Vitamin D**: Vitamin D is involved in the regulation of **immune cell differentiation** and activity, particularly macrophages and T cells. It plays a role in modulating inflammatory responses and promoting immune tolerance, which helps prevent autoimmune reactions. Although the exact mechanisms of vitamin D in fish immunity are not fully understood, studies suggest that it may enhance resistance to bacterial and viral infections. Vitamin D also supports the health of the skeletal system, which indirectly influences fish resilience and overall fitness.
- **Vitamin E**: As a potent **lipid-soluble antioxidant**, vitamin E helps protect cell membranes from oxidative damage, which is critical for the survival and function of immune cells. It also enhances the fish's immune response to vaccines by improving the activity of B cells and T cells. Vitamin E is particularly important in reducing oxidative stress in fish exposed to environmental stressors or high stocking densities. Inadequate vitamin E levels can lead to weakened immune responses and increased vulnerability to diseases.

3.4 Minerals

Minerals are critical for the proper functioning of immune cells and the overall immune response in fish. Key minerals involved in immune function include zinc, selenium, and iron.

- **Zinc**: Zinc is essential for the **development** and **function** of immune cells, including T cells, B cells, and phagocytes. It also plays a role in maintaining the integrity of skin and mucosal barriers, which are critical for preventing pathogen entry. Zinc deficiency can impair immune responses, leading to increased susceptibility to infections. Adequate zinc levels in fish diets have been shown to enhance resistance to bacterial, viral, and parasitic infections.
- **Selenium**: Selenium functions as a cofactor for antioxidant enzymes, such as **glutathione peroxidase**, which helps reduce oxidative stress in immune cells. By neutralizing harmful free radicals, selenium improves immune cell function and enhances fish resilience to infections. Studies have demonstrated that selenium supplementation in aquafeeds can improve survival rates in fish exposed to stressful conditions or disease outbreaks.
- **Iron**: Iron is necessary for the proliferation and differentiation of immune cells, particularly during the response to infection. However,

iron must be carefully regulated, as excess iron can promote the growth of pathogens and lead to oxidative stress. Fish require balanced levels of iron to support immune function while minimizing the risk of pathogen overgrowth. Iron supplementation has been shown to improve the immune response in fish, but over-supplementation can lead to negative health outcomes.

Nutritional strategies that optimize the intake of key proteins, lipids, vitamins, and minerals can significantly enhance the immune function of fish, contributing to better disease resistance and overall health. By focusing on these nutrients, aquafeed formulations can help support the immune systems of fish in intensive aquaculture settings, reducing the reliance on antibiotics and other chemical treatments while improving sustainability and fish welfare.

4. Functional Additives in Immunonutrition

Beyond the traditional nutrients, functional feed additives have emerged as vital tools in enhancing the immune responses of fish in aquaculture. These additives, such as probiotics, prebiotics, immunostimulants, and phytogenic compounds, serve to improve fish health, strengthen disease resistance, and reduce the need for chemical treatments like antibiotics. The use of these functional ingredients in immunonutrition has gained attention for its ability to promote sustainable aquaculture practices while enhancing the overall well-being of farmed fish.

4.1 Probiotics and Prebiotics

Probiotics and prebiotics are two of the most widely studied functional additives in aquaculture due to their ability to improve gut health and boost immune responses. The fish gut microbiome plays a crucial role in immune regulation, and the incorporation of these additives helps maintain a healthy microbial balance, which in turn supports disease resistance.

- **Probiotics**: Probiotics are live beneficial bacteria or microorganisms that colonize the gastrointestinal (GI) tract of fish, promoting a balanced gut microbiota. Probiotics enhance immune function by stimulating the production of **antimicrobial compounds** and **short-chain fatty acids (SCFAs)**, which lower gut pH and inhibit the growth of harmful pathogens. Probiotics also enhance the barrier function of the intestinal lining, reducing the risk of pathogen translocation from the gut into the bloodstream. Certain strains of **Lactobacillus, Bacillus**, and **Enterococcus** have been used in aquafeeds to promote immune cell activation, stimulate the production of **antibodies**, and increase the expression of **cytokines**, leading to improved fish resistance to bacterial, viral, and parasitic infections.

- **Prebiotics**: Prebiotics are **non-digestible carbohydrates** or fibers that selectively stimulate the growth and activity of beneficial gut bacteria, indirectly supporting the immune system. Common prebiotics used in aquaculture include **fructooligosaccharides (FOS), mannanoligosaccharides (MOS)**, and **inulin**. By promoting a healthy gut microbiota, prebiotics help strengthen the fish's natural defenses. Prebiotics also improve gut health by increasing the production of **butyrate**, an SCFA that supports intestinal cell integrity and has anti-inflammatory properties. This can lead to better nutrient absorption, enhanced growth, and a more resilient immune system.

Together, probiotics and prebiotics create a synergistic effect, often referred to as synbiotics, that optimizes gut health and immune function, making fish more resilient to infections and stressful farming conditions.

4.2 Immunostimulants

Immunostimulants are compounds that enhance the activity of the immune system, either by boosting the innate immune response or by enhancing the adaptive immune system. They are commonly used in aquaculture as dietary supplements to improve fish health and disease resistance. Immunostimulants can be derived from natural sources such as yeast, fungi, algae, and plants, or they can be synthetically produced. Some of the most effective immunostimulants include beta-glucans and nucleotides.

- **Beta-glucans**: Beta-glucans are **polysaccharides** found in the cell walls of yeast, fungi, and certain grains. They are well known for their ability to stimulate the immune system by activating **macrophages, natural killer (NK) cells,** and **neutrophils**. These immune cells play a key role in recognizing and eliminating pathogens. Beta-glucans also enhance the production of **cytokines**, which are signaling molecules that regulate the immune response. In fish, beta-glucan supplementation has been shown to improve resistance to bacterial infections like **Aeromonas** and **Vibrio**, as well as viral and parasitic infections. They are commonly added to fish diets to enhance non-specific immunity, especially in situations where fish are exposed to stressful conditions or high disease pressure.
- **Nucleotides**: Nucleotides are the building blocks of **DNA** and **RNA** and are essential for cellular replication, repair, and immune function. While fish can synthesize nucleotides, their demand increases during periods of rapid growth, stress, or disease. Dietary supplementation of nucleotides supports the rapid proliferation of **immune cells** and promotes intestinal health by enhancing the repair of damaged tissues. Nucleotides also

improve the fish's response to vaccinations by enhancing the production of **antibodies**. Research has shown that nucleotide-supplemented diets improve resistance to viral and bacterial infections, reduce recovery time from illness, and enhance the overall immune response in fish.

4.3 Phytogenics

Phytogenics, also known as plant-based compounds, are natural substances derived from herbs, spices, and other plants. These compounds have gained popularity in aquaculture due to their immunomodulatory, antimicrobial, and antioxidant properties. Phytogenics are used in aquafeeds to support fish health, improve immune function, and reduce the incidence of disease. Some commonly used phytogenic compounds include essential oils, polyphenols, and saponins.

- **Essential Oils**: Essential oils are volatile compounds extracted from plants like **thyme, oregano, garlic,** and **eucalyptus**. These oils possess **antimicrobial** properties, helping to inhibit the growth of pathogenic bacteria in the gut. Essential oils also have **anti-inflammatory** and **immunostimulatory** effects, enhancing the activity of immune cells and reducing excessive inflammation that can occur during infections. For example, **garlic essential oil** has been shown to improve the immune response in fish by increasing the activity of macrophages and lymphocytes, leading to better resistance to parasitic infections like *Ichthyophthirius multifiliis*. Similarly, **oregano oil** has been found to reduce the severity of bacterial infections and support overall gut health.
- **Polyphenols**: Polyphenols are plant-derived compounds with strong **antioxidant** properties, protecting immune cells from oxidative damage caused by reactive oxygen species (ROS). Polyphenols, such as **flavonoids, tannins**, and **phenolic acids**, also have **anti-inflammatory** and **antimicrobial** effects. In fish, polyphenol supplementation has been shown to improve immune responses by enhancing the activity of phagocytic cells, increasing the production of antibodies, and modulating cytokine expression. Polyphenols from plants like **green tea** and **turmeric** are particularly effective in promoting fish health and disease resistance.
- **Saponins**: Saponins are naturally occurring compounds found in plants like **quillaja** and **yucca**. They have **immunomodulatory** and **antiparasitic** properties, making them valuable additives in aquafeeds. Saponins enhance the fish's innate immune response by activating macrophages and other immune cells. They also help reduce the load of internal parasites by disrupting the cell membranes of pathogens.

In addition to their immune-enhancing properties, saponins have been found to improve intestinal health by reducing gut inflammation and promoting a balanced microbial community.

The use of functional additives in aquafeeds, such as probiotics, prebiotics, immunostimulants, and phytogenics, plays a critical role in enhancing the immune system of fish and promoting disease resistance. These additives work synergistically with traditional nutrients to improve gut health, boost immune responses, and protect fish from a wide range of pathogens. As aquaculture continues to expand, the integration of these functional additives into feed formulations will become increasingly important in promoting sustainable and healthy fish farming practices. By reducing the reliance on antibiotics and other chemical treatments, functional additives offer a natural and effective way to enhance fish health, improve growth performance, and support the long-term sustainability of the aquaculture industry.

5. Benefits of Immunonutrition in Aquaculture

The integration of immunonutrition into aquaculture practices offers multiple advantages that align with the goals of sustainable, healthy, and productive fish farming. By strategically enhancing immune function through diet, the aquaculture industry can achieve significant benefits in terms of disease resistance, growth performance, and environmental sustainability. Below are some key advantages of immunonutrition:

5.1 Disease Resistance

One of the most critical benefits of immunonutrition is improved disease resistance in farmed fish. By providing the right nutrients and functional additives, the immune system is strengthened, allowing fish to mount more effective responses against pathogens. Fish with enhanced immune systems are better able to fend off infections caused by bacteria, viruses, parasites, and fungi. This not only reduces mortality rates during disease outbreaks but also minimizes the spread of diseases within fish populations. In aquaculture settings where disease outbreaks can lead to devastating economic losses, immunonutrition serves as a preventive measure. Enhancing the natural defense mechanisms of fish reduces the severity of infections and helps farmers maintain healthy stock, which directly translates to more stable production and profitability.

5.2 Reduced Antibiotic Use

The overuse of antibiotics in aquaculture has been a significant concern, contributing to the development of antibiotic-resistant bacteria and the presence of drug residues in fish products. Immunonutrition offers a natural alternative by boosting the immune systems of fish, thereby reducing their reliance on antibiotics and other chemical treatments for disease prevention and control. Healthier fish with stronger immune systems are less susceptible to infections, leading to fewer antibiotic treatments being needed. This reduction in antibiotic use has a positive impact on both human health and the environment. By minimizing the presence of antibiotics in fish products, the risk of antibiotic-resistant pathogens entering the food chain is lowered. Furthermore, reduced chemical use helps prevent environmental contamination, promoting more sustainable aquaculture practices.

5.3 Improved Growth and Survival

Fish that are healthier and less burdened by disease tend to grow more efficiently. Immunonutrition supports optimal growth by enhancing gut health, nutrient absorption, and immune function. Nutrients such as specific amino acids, fatty acids, and vitamins not only help improve immune responses but also promote overall growth and development. A robust immune system allows fish to allocate more energy toward growth and reproduction rather than fighting infections. This results in better feed conversion ratios (FCR) and higher survival rates, both of which are critical for maximizing production efficiency in aquaculture. In many cases, farms that invest in immunonutritional strategies see tangible improvements in fish growth, survival, and profitability.

5.4 Sustainability

Sustainability is a pressing concern in modern aquaculture, as the industry seeks to balance increased production with minimal environmental impact. Immunonutrition contributes to the goal of sustainable aquaculture by promoting natural disease resistance, thereby reducing the reliance on synthetic chemicals like antibiotics, antifungals, and antiparasitics. By focusing on immune health through nutrition, the aquaculture industry can lower its ecological footprint. Healthier fish require fewer chemical interventions, which leads to less contamination of surrounding water bodies and ecosystems. Additionally, immunonutrition aligns with consumer demand for eco-friendly and antibiotic-free seafood products, enhancing the marketability of farmed fish.

6. Challenges and Future Prospects

While the benefits of immunonutrition are well-documented, there are several challenges that limit its widespread adoption. Addressing these challenges is essential for optimizing the application of immunonutrition in aquaculture and unlocking its full potential.

6.1 Cost

The incorporation of high-quality ingredients and functional additives into aquafeeds can be costly, particularly when using specialized nutrients like omega-3 fatty acids, immunostimulants, and phytogenic compounds. These ingredients often come with a higher price tag than conventional feed ingredients, which can be a barrier for small- to medium-sized producers operating on tight margins. To address the cost barrier, ongoing research is focused on developing more cost-effective formulations that still provide immune-enhancing benefits. As the demand for immunonutritional products grows, economies of scale may also help reduce costs over time, making these feeds more accessible to a wider range of producers.

6.2 Inconsistent Results

Another challenge with immunonutrition is the variability in results, which can depend on factors such as fish species, environmental conditions, and pathogen load. What works well for one species under certain farming conditions may not be as effective for another species or in a different environment. In some cases, immunonutritional strategies may produce inconsistent results in terms of disease resistance and growth performance. To overcome this, more species-specific and condition-specific research is needed to fine-tune immunonutrition approaches. This includes identifying optimal nutrient levels and functional additives for different fish species and farming systems. Precision nutrition tools, which use data to tailor feeds to the specific needs of individual fish, may also help enhance the consistency of results.

6.3 Research Gaps

While significant progress has been made in understanding the role of nutrition in fish immune function, there are still many gaps in the research. For example, the mechanisms by which certain nutrients and additives modulate the immune system are not fully understood. Additionally, there is a need for long-term studies that evaluate the effects of immunonutrition over the entire production cycle, from juvenile stages to market size. Future research will likely focus on elucidating the molecular pathways involved in immunonutrition, as well as exploring the potential of novel feed ingredients, such as algae, insect-based proteins, and synthetic amino acids, to enhance fish immunity. Advances in biotechnology and genomics may also lead to the development of genetically enhanced feed ingredients that provide more targeted immune support.

6.4 Future Prospects

The future of immunonutrition in aquaculture looks promising, particularly with the development of precision nutrition tools, which use data and technology to deliver tailored feed formulations that meet the specific immune and nutritional needs of fish. Innovations in feed ingredient sourcing, such as using insect meal and microalgae, are also expected to reduce costs while providing sustainable sources of immune-enhancing nutrients. Genetically enhanced feed ingredients are another exciting frontier. Advances in genetic engineering could allow for the production of feed ingredients that are specifically designed to modulate immune function, enhance growth, and improve overall health. These innovations have the potential to revolutionize the way fish are fed and managed, ultimately making immunonutrition more effective and accessible.

7. Conclusion

Immunonutrition is a powerful tool for improving disease resistance, fish health, and sustainability in aquaculture. By incorporating key nutrients and functional additives into aquafeeds, producers can enhance the natural defense mechanisms of fish, reducing the need for antibiotics and other chemical treatments. Immunonutrition supports healthier, faster-growing fish that are more resilient to disease, which in turn boosts production efficiency and reduces environmental impacts. As the industry moves toward more sustainable practices, immunonutrition will play an increasingly important role in maintaining the health and productivity of farmed fish. Despite challenges related to cost and variability, ongoing research and technological advancements are poised to make immunonutrition an essential component of aquaculture in the future. By embracing immunonutritional strategies, producers can meet the growing demand for safe, healthy, and sustainable seafood, benefiting both the industry and consumers.

References

Ahmadifar, E., Sheikhzadeh, N., Razeghi Mansour, M., & Dadar, M. (2019). Probiotics for fish: A review of current knowledge and future perspectives. Journal of Animal Physiology and Animal Nutrition, 103(2), 152-162.

Akrami, R., Iri, Y., & Rostami, H. (2020). The use of immunostimulants in aquaculture: A review. Reviews in Aquaculture, 12(1), 189-210.

Dadar, M., Dhama, K., Vakharia, V. N., Hoseinifar, S. H., Karthik, K., Tiwari, R., & Singh, S. D. (2017). Advances in aquaculture vaccines against fish pathogens: Global status and trends. Reviews in Fisheries Science & Aquaculture, 25(3), 184-217.

De Schryver, P., Defoirdt, T., & Sorgeloos, P. (2014). Early mortality syndrome outbreaks: A microbial management issue in shrimp farming? PLoS Pathogens, 10(3), e1003919.

Defoirdt, T., Sorgeloos, P., & Bossier, P. (2011). Alternatives to antibiotics for the control of bacterial disease in aquaculture. Current Opinion in Microbiology, 14(3), 251-258.

Esteban, M. Á., & Meseguer, J. (2009). Fish nutrition and immune system: An overview. In B. Austin & S. Austin (Eds.), Fish Defenses: Immunology (pp. 229-264). Science Publishers.

Gupta, S. K., Pal, A. K., Sahu, N. P., Dalvi, R., Kumar, V., & Mukherjee, S. C. (2007). Dietary microbial levan enhances cellular non-specific immunity and survival of common carp (Cyprinus carpio) juveniles. Fish & Shellfish Immunology, 22(4), 477-486.

Hoseinifar, S. H., Esteban, M. Á., Cuesta, A., & Sun, Y. Z. (2015). Prebiotics and fish immune response: A review of current knowledge and future perspectives. Reviews in Fisheries Science & Aquaculture, 23(4), 315-328.

Hoseinifar, S. H., Ringø, E., Shenavar Masouleh, A., & Esteban, M. Á. (2016). Probiotic, prebiotic and synbiotic supplements in sturgeon aquaculture: A review. Reviews in Aquaculture, 8(1), 89-102.

Kiron, V. (2012). Fish immune system and its nutritional modulation for preventive health care. Animal Feed Science and Technology, 173(1-2), 111-133.

Magnadóttir, B. (2010). Immunological control of fish diseases. Marine Biotechnology, 12(4), 361-379.

Naylor, R. L., Goldburg, R. J., Primavera, J. H., Kautsky, N., Beveridge, M. C., Clay, J., ... & Troell, M. (2000). Effect of aquaculture on world fish supplies. Nature, 405(6790), 1017-1024.

Oliva-Teles, A. (2012). Nutrition and health of aquaculture fish. Journal of Fish Diseases, 35(2), 83-108.

Paray, B. A., Al-Sadoon, M. K., & Al-Misned, F. (2021). Dietary plant extracts and fish immunity: A review. Reviews in Aquaculture, 13(2), 896-911.

Puangkaew, J., Kiron, V., Satoh, S., Watanabe, T., & Takeuchi, T. (2005). Dietary nucleotide supplementation enhances disease resistance and digestive enzyme activities in Japanese flounder, Paralichthys olivaceus. Aquaculture, 244(1-4), 115-121.

Rawling, M. D., Merrifield, D. L., & Davies, S. J. (2009). Preliminary assessment of dietary supplementation of Sangrovit® on red tilapia (Oreochromis niloticus) growth performance and health. Aquaculture, 294(1-2), 118-122.

Reyes-Becerril, M., Tovar-Ramírez, D., Ascencio-Valle, F., & Gracia-López, V. (2011). Dietary immunostimulants: Enhancing immune responses in fish. African Journal of Biotechnology, 10(84), 19631-19645.

Ringø, E., & Song, S. K. (2016). Application of dietary supplements (synbiotics and probiotics in combination with plant products and β-glucans) in aquaculture. Aquaculture Nutrition, 22(1), 4-24.

Ringo, E., Olsen, R. E., Gifstad, T. Ø., Dalmo, R. A., Amlund, H., Hemre, G. I., & Bakke, A. M. (2010). Prebiotics in aquaculture: A review. Aquaculture Nutrition, 16(2), 117-136.

Shinn, A. P., Pratoomyot, J., Griffiths, D., Trong, T. Q., Vu, N. V., Jiravanichpaisal, P., & Briggs, M. (2018). Asian shrimp aquaculture in the post-antibiotic era: Solutions to bacterial disease management in farmed shrimp. Reviews in Aquaculture, 10(2), 208-233.

Turchini, G. M., Torstensen, B. E., & Ng, W. K. (2009). Fish oil replacement in finfish nutrition. Reviews in Aquaculture, 1(1), 10-57.

Wang, Y., & Li, J. (2014). Microbial immunonutrition in fish. Fish & Shellfish Immunology, 40(2), 255-258.

Zhang, J., Liu, Y., Tian, L., Yang, H., Liang, G., & Xu, D. (2013). Effects of dietary nucleotide-rich yeast extract on growth, non-specific immunity, antioxidant capacity and disease resistance of juvenile Pacific white shrimp (Litopenaeus vannamei). Fish & Shellfish Immunology, 34(5), 1128-1135.

11

Nutritional Strategies for Stress Management in Aquaculture

1. Introduction

Aquaculture is one of the most rapidly expanding sectors in food production, making significant contributions to global food security and the economy. As natural fisheries resources are depleted, aquaculture has become an essential source of protein and employment for millions worldwide. With an increasing demand for seafood, the intensification of aquaculture systems has become necessary. This intensification, characterized by high stocking densities, increased fish production, and automation, has introduced a range of challenges that negatively impact the health and well-being of farmed fish and other aquatic species.

1.1 Stress in Aquaculture Systems

Stress is an inevitable part of aquaculture, particularly in intensive farming systems where fish are exposed to multiple environmental, biological, and anthropogenic stressors. Stress refers to a physiological condition where the normal homeostasis of an organism is disrupted due to external stimuli. In aquaculture, stress can arise from various factors, including changes in water quality parameters (such as dissolved oxygen, temperature, and pH), overcrowding, transportation, handling, and exposure to pathogens. The physiological response to stress can lead to a cascade of negative outcomes, including suppressed immunity, impaired growth, decreased reproductive success, and increased susceptibility to disease. If not adequately managed, stress can cause significant economic losses, affecting the sustainability and profitability of aquaculture operations.

1.2 Effects of Stress on Fish Health

When fish are exposed to stress, their neuroendocrine system is activated, resulting in the release of stress hormones such as cortisol and catecholamines. These hormones trigger a series of physiological and metabolic changes designed to help the organism cope with the stressor. While short-term stress responses are adaptive and help fish survive immediate threats, chronic or prolonged stress can lead to deleterious effects. Key consequences of chronic stress in aquaculture include:

- **Immune Suppression**: Prolonged stress suppresses the immune system, making fish more vulnerable to bacterial, viral, and parasitic infections.
- **Reduced Growth and Feed Efficiency**: Stress often leads to a decrease in appetite and feed intake, resulting in slower growth and reduced feed conversion efficiency. Energy that would typically be used for growth is redirected towards coping with stress, further hindering performance.
- **Poor Reproductive Success**: Chronic stress can disrupt the endocrine system, impairing reproductive processes such as gamete development, spawning, and fertilization.
- **Increased Mortality**: Fish under constant stress have a higher risk of mortality due to weakened physiological defenses and increased susceptibility to diseases and environmental challenges.

1.3 The Role of Nutrition in Stress Mitigation

In recent years, nutritional strategies have gained significant attention as a key method for managing stress in aquaculture systems. Nutrient supplementation and feed formulation play an essential role in promoting fish health and enhancing their ability to cope with stressors. Proper nutrition supports metabolic processes, immune function, and antioxidative defense mechanisms that help fish resist and recover from stress. For example, specific nutrients such as amino acids, fatty acids, vitamins, and minerals can modulate the immune system, promote antioxidant activity, and enhance energy metabolism, all of which are crucial during stressful conditions.

1.4 Nutritional Interventions for Stress Alleviation

Nutritional interventions aim to strengthen the physiological resilience of fish, improving their capacity to handle various stressors and reducing the negative impacts of stress on growth and survival. By incorporating stress-mitigating nutrients into aquafeeds, fish farmers can enhance the overall performance of cultured species while reducing the need for therapeutic treatments like antibiotics or chemicals. Several nutrients and functional feed additives have been identified as effective in reducing stress-related effects:

- **Proteins and Amino Acids**: These are crucial for muscle maintenance, immune function, and repair of damaged tissues during stress. Certain amino acids, such as glutamine, arginine, and taurine, are particularly important for supporting the immune response and managing metabolic changes during stress.

- **Lipids and Fatty Acids**: Omega-3 and omega-6 fatty acids help regulate inflammation and oxidative stress. Supplementing aquafeeds with omega-3 fatty acids, especially EPA and DHA, has been shown to improve stress tolerance and reduce inflammatory responses in fish.
- **Vitamins and Minerals**: Antioxidant vitamins such as vitamin C and E, as well as trace minerals like zinc and selenium, help mitigate oxidative stress and maintain immune function. These nutrients play key roles in protecting cells from damage caused by stress-induced reactive oxygen species (ROS).
- **Functional Feed Additives**: Probiotics, prebiotics, phytogenics (plant-based additives), and immunostimulants (such as beta-glucans) are increasingly used to support gut health, boost immune function, and promote resilience to stress in aquaculture species.

2. Understanding Stress in Aquaculture

Stress in aquaculture involves complex physiological processes triggered by environmental, biological, and handling-related factors. When fish encounter stress, their bodies initiate a "fight or flight" response, marked by the release of stress hormones that prepare them to cope with the challenge. However, prolonged exposure to stress can compromise their health, leading to immune suppression, metabolic disturbances, and increased susceptibility to diseases. Understanding the causes and consequences of stress is essential for implementing effective nutritional strategies to manage it.

2.1 Environmental Stress

Environmental stress is one of the most common forms of stress in aquaculture and can arise from changes in water quality, including fluctuations in temperature, salinity, dissolved oxygen, ammonia, and pH. For instance, sudden changes in water temperature can disrupt metabolic processes, while low dissolved oxygen levels can lead to hypoxia, causing oxidative stress and impairing fish performance. Poor water quality can also exacerbate the effects of other stressors, making fish more vulnerable to pathogens and diseases.

2.2 Biological Stress

Biological stressors include the presence of pathogens such as bacteria, viruses, fungi, and parasites that can trigger immune responses in fish. The competition for resources among fish, particularly in high-density aquaculture systems, can also contribute to biological stress. When fish are stressed, their immune systems become compromised, making them more susceptible to diseases caused by these biological agents. Disease outbreaks in aquaculture systems can result in significant economic losses due to increased mortality, reduced growth, and the need for medical treatments.

2.3 Handling Stress

Handling stress occurs when fish are exposed to physical manipulation during activities such as transportation, grading, sorting, and harvesting. These procedures are unavoidable in aquaculture operations but can cause significant physiological stress in fish. Handling stress is often associated with physical injuries, increased metabolic demand, and elevated cortisol levels, all of which contribute to immune suppression and poor growth performance. Proper handling techniques and stress-reducing measures are essential for minimizing the negative impacts of handling on fish health.

2.4 The Stress Response

When fish experience stress, their hypothalamus-pituitary-interrenal (HPI) axis is activated, resulting in the release of cortisol, the primary stress hormone in fish. Cortisol plays a key role in regulating metabolism, immune function, and osmoregulation, but prolonged exposure to high cortisol levels can have negative effects, including:

- **Immune Suppression**: Chronic stress leads to a decrease in the activity of immune cells such as macrophages and lymphocytes, reducing the fish's ability to fight off infections.
- **Metabolic Imbalance**: Stress diverts energy away from growth and reproduction toward survival mechanisms. As a result, stressed fish often show reduced feed intake and poor growth performance.
- **Oxidative Stress**: Stress increases the production of ROS, which can damage cells and tissues. If not neutralized by antioxidants, oxidative stress can lead to cell death and organ dysfunction.

Each type of stressor can exacerbate the effects of the others, creating a cycle of immune suppression, metabolic imbalance, and oxidative damage that further compromises the health of fish. By understanding these stress responses, nutritional strategies can be designed to mitigate the adverse effects of stress and support fish health during stressful conditions.

3. Role of Proteins and Amino Acids in Stress Management

Proper nutrition is crucial for managing stress in aquaculture, and proteins and amino acids play central roles in supporting fish health during stressful conditions. Understanding how these macronutrients contribute to stress resilience can help aquaculture producers formulate effective feeds that enhance fish welfare and performance.

3.1 Proteins

Proteins are essential macromolecules that perform a myriad of functions within the body, including structural support, enzymatic activity, and immune response regulation. In the context of stress management, the role of dietary protein becomes increasingly significant for several reasons:

- **Growth and Tissue Repair**: During stress, fish often experience a reduction in feed intake, which can lead to protein depletion as the body begins to utilize its protein stores for energy and metabolic functions. Providing adequate dietary protein helps mitigate these effects, promoting growth and repair of tissues that may be damaged due to stress.
- **Synthesis of Stress Proteins**: Stressful conditions induce the synthesis of heat shock proteins (HSPs), which act as molecular chaperones, protecting cellular structures and assisting in the refolding of denatured proteins. HSPs are crucial for maintaining cellular integrity under stress, and sufficient protein intake is necessary for their production.
- **Immune Support**: Proteins are vital for the synthesis of antibodies and cytokines, which are essential components of the immune response. When fish are stressed, their immune systems can become compromised, making them more susceptible to diseases. Ensuring a high-quality protein source in the diet supports immune function and resilience against pathogens.
- **Energy Production**: Although carbohydrates and fats are the primary energy sources, proteins can be catabolized for energy when other sources are insufficient. This process becomes crucial during stress, when energy demands increase, and other nutrients may not be readily available.

3.2 Amino Acids

Amino acids, the building blocks of proteins, play critical roles in stress management beyond their structural functions. Certain amino acids are particularly effective in supporting the immune system, enhancing antioxidant defenses, and facilitating recovery during stress.

- **Glutamine**
  - **Role**: Glutamine is an important energy source for rapidly dividing cells, including intestinal and immune cells. It supports the integrity of the gut barrier, which is crucial for maintaining optimal digestion and nutrient absorption.

    - **Impact on Stress**: During stressful conditions, glutamine supplementation can help maintain gut health and immune function, reducing the risk of intestinal permeability and subsequent infections.

- **Arginine**
    - **Role**: Arginine is a precursor for nitric oxide (NO), a signaling molecule that plays roles in vascular regulation, neurotransmission, and immune responses.
    - **Impact on Stress**: Supplementation with arginine has been shown to enhance immune responses, promoting the activity of macrophages and T cells. This is particularly beneficial during stress, as it helps improve fish resilience and reduces mortality rates.
- **Taurine**
    - **Role**: Taurine is an amino acid known for its antioxidant properties and ability to regulate calcium levels within cells. It plays a role in osmoregulation, which is especially important for fish exposed to fluctuating salinity levels.
    - **Impact on Stress**: Taurine supplementation can help mitigate oxidative stress, modulate immune responses, and improve overall stress tolerance in aquaculture species.

4. Lipids and Fatty Acids in Stress Mitigation

Lipids, particularly polyunsaturated fatty acids (PUFAs), are vital components of cell membranes, energy storage, and hormone production. They also play a significant role in modulating immune and inflammatory responses, making them essential for stress management in aquaculture.

4.1 Omega-3 Fatty Acids (EPA and DHA)

Omega-3 fatty acids, especially eicosapentaenoic acid (EPA) and docosahexaenoic acid (DHA), are known for their health benefits in both aquatic and terrestrial animals.

- **Anti-inflammatory Properties**: Omega-3 fatty acids help to reduce the production of pro-inflammatory cytokines, which are implicated in stress-related tissue damage. By modulating the inflammatory response, these fatty acids protect tissues from excessive damage during stress events.
- **Enhancing Stress Resilience**: Supplementing aquafeeds with omega-3 fatty acids has been shown to improve fish resilience to environmental stressors, such as temperature fluctuations and hypoxia. They support metabolic processes during stress and enhance overall fish performance.

4.2 Omega-6 Fatty Acids

While omega-6 fatty acids are essential for immune responses and inflammation, maintaining a balanced ratio of omega-3 to omega-6 fatty acids is crucial for minimizing stress-related inflammation.

- **Role in Inflammation**: Omega-6 fatty acids, such as arachidonic acid, are precursors to pro-inflammatory eicosanoids. While these compounds are necessary for a robust immune response, excessive levels can lead to prolonged inflammation, potentially exacerbating stress.
- **Balancing Omega-3 and Omega-6**: Ensuring a proper balance between omega-3 and omega-6 fatty acids in aquafeeds can help mitigate the negative impacts of stress on fish health. An optimal ratio supports effective immune responses while minimizing excessive inflammation.

5. Vitamins and Minerals for Stress Alleviation

Vitamins and minerals are critical cofactors in numerous metabolic and antioxidant pathways, playing vital roles in stress management.

5.1 Vitamin C

Vitamin C (ascorbic acid) is a potent antioxidant that plays several key roles during stressful conditions:

- **Antioxidant Activity**: Vitamin C neutralizes reactive oxygen species (ROS) generated during stress, helping to protect cells from oxidative damage. It can also regenerate other antioxidants, enhancing overall antioxidant capacity.
- **Immune Function and Wound Healing**: Vitamin C is essential for collagen synthesis, which is crucial for wound healing. Adequate vitamin C levels support immune function, reducing the risk of disease during stressful periods.
- **Supplementation Benefits**: Research has shown that vitamin C supplementation can improve survival rates and reduce oxidative damage in fish exposed to stressors such as handling and disease.

5.2 Vitamin E

Vitamin E (tocopherol) is another critical antioxidant that plays a protective role during stress:

- **Cell Membrane Protection**: Vitamin E protects cell membranes from lipid peroxidation, which can occur during oxidative stress. This protection helps maintain cellular integrity and function.

- **Synergistic Effects with Vitamin C**: When combined with vitamin C, vitamin E enhances the antioxidant defense system of fish, reducing the negative impacts of stress on immune function and overall health.

5.3 Vitamin D

Vitamin D is crucial for calcium homeostasis and immune modulation:

- **Immune Modulation**: In fish, vitamin D enhances macrophage function and supports T cell activation, contributing to improved stress resilience.
- **Calcium Regulation**: Proper calcium levels are essential for muscle contraction, nerve function, and overall physiological processes, making vitamin D supplementation vital during stressful conditions.

5.4 Zinc

Zinc is an essential trace mineral involved in several physiological processes:

- **Immune System Support**: Zinc is crucial for immune cell development and function. It acts as a cofactor for various enzymes involved in immune responses.
- **Antioxidant Properties**: Zinc exhibits antioxidant properties, protecting fish from oxidative stress. Adequate zinc levels can enhance the immune response and reduce mortality during disease outbreaks, particularly in stress-prone aquaculture systems.

5.5 Selenium

Selenium is an essential trace mineral that serves as a cofactor for the enzyme glutathione peroxidase, which plays a vital role in cellular antioxidant defense:

- **Oxidative Damage Protection**: Selenium supplementation can help protect cells from oxidative damage caused by stress. Adequate selenium levels are crucial for maintaining immune function and overall health.
- **Preventing Deficiencies**: Selenium deficiency can lead to increased susceptibility to stress and disease, highlighting the importance of ensuring appropriate selenium levels in aquafeeds.

6. Functional Additives for Stress Management

The use of functional additives in aquaculture feeds has gained significant attention for their potential to enhance fish health and resilience against stress. These additives can modulate physiological responses and improve overall performance in aquaculture species. This section delves deeper into the key functional additives that play a role in stress management.

6.1 Probiotics

Probiotics are live microorganisms that, when administered in adequate amounts, confer health benefits to the host, particularly through their positive effects on gut health and immune function.

- **Mechanisms of Action**: Probiotics can improve gut health by restoring the balance of gut microbiota, enhancing nutrient absorption, and producing beneficial metabolites. They compete with pathogenic bacteria for space and nutrients, thereby inhibiting pathogen colonization and growth. Moreover, probiotics can produce antimicrobial compounds, such as bacteriocins, that further protect the host from infections.
- **Stress Resilience**: Probiotic supplementation has been shown to enhance stress resilience in fish by modulating immune responses and reducing the negative effects of stressors. For example, studies have indicated that probiotics can decrease cortisol levels and improve the overall health of fish during handling and environmental stressors.
- **Specific Strains**: Certain strains of probiotics, such as Lactobacillus and Bacillus species, have demonstrated beneficial effects in various fish species, leading to improved growth rates, enhanced immune responses, and reduced mortality during stress.

6.2 Prebiotics

Prebiotics are non-digestible food components, primarily carbohydrates that selectively stimulate the growth and activity of beneficial gut bacteria.

- **Impact on Gut Health**: Prebiotics serve as substrates for beneficial microorganisms in the gut, promoting their growth and activity. This process supports a healthy gut microbiome, which is crucial for nutrient absorption and immune function.
- **Enhanced Immune Function**: By fostering a balanced gut microbiota, prebiotics can enhance the immune system's ability to respond to pathogens, ultimately improving fish performance under stressful conditions. For instance, the use of oligosaccharides as prebiotics has been associated with enhanced immune responses and resistance to disease.
- **Common Sources**: Common prebiotic sources in aquaculture diets include inulin, fructooligosaccharides (FOS), and mannanoligosaccharides (MOS), which can be derived from various plant materials.

6.3 Phytogenics

Phytogenics are bioactive plant-based compounds that possess a range of beneficial properties, including antioxidant, anti-inflammatory, and immunomodulatory effects.

- **Antioxidant Properties**: Plant extracts containing polyphenols, flavonoids, and essential oils can help reduce oxidative stress in fish. For example, garlic (Allium sativum) and thyme (Thymus vulgaris) extracts have shown promise in scavenging free radicals and enhancing antioxidant defenses.
- **Immunomodulation**: Phytogenic compounds can modulate immune responses by enhancing the activity of immune cells, such as macrophages and lymphocytes. This modulation is crucial for improving fish resilience to environmental stressors and pathogens.
- **Examples**: Essential oils derived from herbs like oregano, rosemary, and clove have been utilized in aquaculture diets to enhance stress tolerance and improve overall health. These plant extracts can also enhance feed palatability, leading to better nutrient intake.

6.4 Beta-Glucans

Beta-glucans are polysaccharides found in the cell walls of yeast and fungi, recognized for their immunostimulatory properties.

- **Mechanism of Action**: Beta-glucans act as immunomodulators by activating various immune cells, including macrophages, neutrophils, and natural killer cells. They enhance phagocytic activity and the production of pro-inflammatory cytokines, thereby improving the immune response of fish.
- **Stress Resilience**: Supplementation with beta-glucans has been linked to increased resistance to infections and improved stress coping mechanisms in aquaculture species. They can also enhance growth performance by reducing the negative impacts of stress on fish metabolism.
- **Sources**: Common sources of beta-glucans include yeast extracts, especially from species such as Saccharomyces cerevisiae, which can be incorporated into aquafeeds for their immune-enhancing effects.

7. Stress Management through Dietary Strategies

Effective dietary strategies are essential for mitigating stress in aquaculture species. By formulating diets that address the specific needs of fish during stressful periods, producers can enhance growth performance and overall health.

- **Optimizing Protein and Energy Ratios**: It is vital to ensure that fish receive adequate protein and energy levels in their diets, especially during stressful situations. Proper ratios can prevent growth suppression and immune dysfunction, enabling fish to maintain their physiological processes.
- **Supplementing with Antioxidants**: Adding antioxidants, such as vitamins C and E, as well as natural antioxidants like carotenoids, can help mitigate oxidative damage associated with stress. These supplements can enhance the antioxidant capacity of fish, protecting cells from oxidative stress.
- **Incorporating Functional Feed Additives**: Including probiotics, prebiotics, phytogenics, and immunostimulants in the diet can further enhance stress resistance and improve overall health and performance. These additives support gut health, enhance immune function, and provide antioxidant protection, helping fish cope with various stressors.

8. Future Prospects and Challenges

While nutritional strategies for stress management hold great promise, several challenges remain in their implementation within the aquaculture industry:

- **Cost Considerations**: The inclusion of high-quality ingredients and functional additives in aquafeeds may increase feed costs. This potential barrier could affect the willingness of aquaculture producers to adopt these strategies, especially in regions with tight profit margins.
- **Species-Specific Responses**: The effectiveness of nutritional strategies may vary based on the species being cultured, environmental conditions, and the specific type of stressor encountered. Future research will be essential to identify the optimal nutritional requirements for different aquaculture species and develop tailored strategies for stress management.
- **Advancements in Biotechnology**: The integration of nutrigenomics and microbiome analysis will pave the way for precision nutrition strategies that are both cost-effective and sustainable. By understanding the specific nutritional needs of different species and their gut microbiota, aquaculture producers can design feeds that optimize stress resilience.

Table: Various nutritional strategies for stress management in aquaculture, including the components involved, their roles, and potential benefits

Nutritional Strategy	Components	Role in Stress Management	Potential Benefits
Proteins and Amino Acids	- High-quality proteins	Support growth, tissue repair, and stress protein synthesis	- Enhanced growth and recovery during stress
	- Glutamine	Energy source for gut and immune cells	- Maintains gut integrity and supports immune function
	- Arginine	Precursor for nitric oxide, enhances immune response	- Improved stress resilience
	- Taurine	Antioxidant properties, osmoregulation	- Better adaptation to salinity changes
Lipids and Fatty Acids	- Omega-3 Fatty Acids (EPA, DHA)	Anti-inflammatory, supports cellular integrity	- Reduced tissue damage during stress
	- Omega-6 Fatty Acids	Essential for immune response	- Balanced immune function
Vitamins	- Vitamin C	Antioxidant, supports collagen synthesis and immune function	- Reduced oxidative damage, improved survival
	- Vitamin E	Protects cell membranes from oxidative stress	- Enhanced immune capacity
	- Vitamin D	Modulates calcium homeostasis and immune responses	- Improved macrophage function
Minerals	- Zinc	Essential for immune cell development	- Enhanced immune response, reduced disease mortality
	- Selenium	Cofactor for antioxidant enzymes	- Supports cellular protection from oxidative stress
Functional Additives	- Probiotics	Enhance gut health and immune responses	- Improved nutrient absorption, stress resilience
	- Prebiotics	Stimulate growth of beneficial gut bacteria	- Enhanced immune function, better overall performance
	- Phytogenics (plant extracts)	Antioxidant and immunomodulatory properties	- Reduced oxidative stress, improved immune responses
	- Beta-Glucans	Immunostimulant, enhances immune cell activity	- Increased resistance to stress and infections

Dietary Strategies	- Optimized protein-energy ratios	Prevents growth suppression and immune dysfunction	- Maintained performance under stress
	- Antioxidant supplementation	Mitigates oxidative damage	- Enhanced stress resistance
	- Inclusion of functional additives	Supports overall health and resilience	- Improved growth rates and reduced disease outbreaks

9. Conclusion

Nutritional strategies play a crucial role in managing stress in aquaculture species. By providing the necessary nutrients and functional additives, aquaculture diets can support immune systems, antioxidant defenses, and overall health during stressful conditions. Implementing these strategies can help reduce disease outbreaks, improve growth and survival rates, and enhance the sustainability of aquaculture practices. As the industry continues to evolve, a focus on nutritional approaches to stress management will be essential for maintaining productivity and ensuring the well-being of farmed fish.

References

Abdel-Tawwab, M., & Monier, M. N. (2018). Dietary L-tryptophan: A potential feed additive to mitigate stress response in farmed fish. Aquaculture, 495, 414-422.

Ahmadifar, E., Fadaei Razi, V., Dargahi, N., et al. (2019). The effects of dietary prebiotics, probiotics, and synbiotics on gut microbiota, stress resistance, and immune response in aquaculture: A review. Fish and Shellfish Immunology, 86, 117-125.

Bai, N., Wang, L., Zhao, X., & Zhang, P. (2020). Effects of dietary vitamin C supplementation on growth performance, immune responses, and oxidative stress resistance in aquatic animals: A review. Aquaculture Nutrition, 26(2), 439-452.

Baruah, K., Sahu, N. P., Pal, A. K., & Debnath, D. (2008). Dietary phytase: An ideal approach for a cost-effective and low-polluting aquafeed. Aquaculture Nutrition, 14(6), 508-519.

Dawood, M. A., Koshio, S., Ishikawa, M., et al. (2017). Effects of dietary supplementation of astaxanthin and vitamin E on the antioxidant response and stress resistance of red sea bream (Pagrus major) exposed to hypoxia. Aquaculture, 477, 104-111.

Dong, X., Xu, H., Yu, Y., & Liang, Y. (2019). Effects of dietary taurine supplementation on growth performance, antioxidant status, and osmoregulation capacity in aquaculture species under environmental stress. Aquaculture Nutrition, 25(5), 1007-1020.

Elumalai, P., Tang, K., Liu, J., et al. (2020). Beneficial effects of dietary probiotics on stress tolerance, gut health, and immunity in aquaculture species: A review. Aquaculture Reports, 18, 100449.

Fasakin, E. A., & Balogun, A. M. (1998). Vitamin C requirements of fish during stressful conditions. Aquaculture Research, 29(1), 25-32.

Hoseinifar, S. H., Esteban, M. Á., Cuesta, A., & Sun, Y. Z. (2015). Prebiotics and fish immune response: A review of current knowledge and future perspectives. Reviews in Fisheries Science & Aquaculture, 23(4), 315-328.

Jia, Z., Xia, W., Wang, S., et al. (2021). Alleviation of stress in aquaculture by nutritional modulation: A review of dietary strategies. Frontiers in Marine Science, 8, 718723.

Kiron, V. (2012). Fish immune system and the benefits of probiotics, prebiotics, and synbiotics in aquaculture: A review. Aquaculture, 412-413, 1-14.
Li, Y., Ma, Y., Jiang, X., et al. (2020). The role of dietary antioxidants in modulating stress response in fish under aquaculture conditions. Aquaculture Nutrition, 26(3), 741-752.
Liu, H., Wang, S., Cai, Y., & Guo, X. (2019). Dietary methionine alleviates oxidative stress in aquaculture species exposed to environmental stressors. Fish and Shellfish Immunology, 87, 379-387.
Manning, F. (2018). Stress reduction in aquaculture through dietary magnesium supplementation. Journal of Applied Aquaculture, 30(1), 25-34.
Martinez-Porchas, M., Martinez-Cordova, L. R., & Ramos-Enriquez, R. (2009). Cortisol and glucose: Reliable indicators of fish stress? Pan-American Journal of Aquatic Sciences, 4(2), 158-178.
Mohapatra, S., Chakraborty, T., Prusty, A. K., et al. (2013). Use of different microbial probiotics in the diet of rohu (Labeo rohita) fingerlings: Effects on growth, nutrient digestibility, and metabolic status. Aquaculture Nutrition, 19(2), 144-153.
Pal, G. K., Raghunathan, M., & George, E. G. (2018). Role of herbal adaptogens in mitigating stress in aquaculture. Aquaculture Nutrition, 24(5), 1345-1360.
Pan, Y., Luo, L., Hu, W., & Wang, W. (2020). Effects of dietary selenium on stress resistance and immune response in aquaculture species. Aquaculture Nutrition, 26(4), 1045-1054.
Pérez-Sánchez, T., Ruiz-Zarzuela, I., & Balcázar, J. L. (2014). Probiotics in aquaculture: A current assessment. Reviews in Aquaculture, 6(2), 133-146.
Saravanan, S., Schrama, J. W., & Verreth, J. A. (2012). Dietary organic acids modulate stress resistance and performance in fish under suboptimal environmental conditions. Fish Physiology and Biochemistry, 38(1), 401-409.
Singh, A. K., & Tiwari, V. K. (2019). Functional feeds and their role in mitigating stress in aquaculture: A review. Aquaculture Research, 50(12), 3563-3573.
Yu, Y., Wang, S., Song, X., et al. (2021). Advances in stress management strategies for improving aquaculture production. Frontiers in Marine Science, 8, 741925.

12

The Role of Omega-3 Fatty Acids in Fish Nutrition and Human Health

1. Introduction

Omega-3 fatty acids are a vital group of polyunsaturated fatty acids (PUFAs) recognized for their complex chemical structure, which is characterized by multiple double bonds within their hydrocarbon chains. This unique arrangement not only influences their physical and biochemical properties but also underpins their essential roles in various biological processes across a wide range of organisms. Omega-3 fatty acids are categorized into three primary types: alpha-linolenic acid (ALA), eicosapentaenoic acid (EPA), and docosahexaenoic acid (DHA). While ALA is predominantly found in plant sources, such as flaxseeds, chia seeds, and walnuts, EPA and DHA are primarily sourced from marine organisms, particularly fish and algae.

1.1 Importance of Omega-3 Fatty Acids

1.1.1 Nutritional Significance in Aquatic Organisms

In fish nutrition, omega-3 fatty acids are indispensable for promoting optimal growth, development, and overall health. These essential nutrients contribute to a multitude of physiological processes, including:

1. **Cellular Structure and Function**: Omega-3 fatty acids are integral components of cell membranes, influencing their fluidity and permeability. This is crucial for maintaining cellular integrity and facilitating the transport of nutrients and signaling molecules across membranes.
2. **Growth and Development**: Adequate dietary levels of EPA and DHA are essential for the growth of fish species. Research has shown that fish fed diets enriched with omega-3 fatty acids exhibit improved growth rates, enhanced feed conversion efficiency, and better muscle development. This is particularly important in aquaculture settings, where maximizing growth and production efficiency is essential for economic viability.

3. **Reproductive Performance**: Omega-3 fatty acids play a significant role in the reproductive health of fish. They contribute to the quality of gametes (sperm and eggs), influencing fertilization success and embryo viability. Studies have demonstrated that dietary inclusion of EPA and DHA can enhance reproductive outcomes, leading to higher larval survival rates and better overall broodstock performance.
4. **Immune Function**: The immune system in fish is often challenged by environmental stressors, pathogens, and changes in water quality. Omega-3 fatty acids bolster immune responses by modulating the activity of immune cells, improving resistance to infections, and reducing the impact of inflammation. This is particularly critical for fish raised in aquaculture, where crowded conditions and stress can increase susceptibility to disease.

1.1.2 Health Benefits in Humans

In addition to their essential roles in fish nutrition, omega-3 fatty acids have garnered considerable attention in human health research. Numerous studies have elucidated their beneficial effects on various aspects of human health, including:

1. **Cardiovascular Health**: Omega-3 fatty acids have been linked to a reduced risk of cardiovascular diseases. They are known to lower triglyceride levels, decrease blood pressure, and inhibit platelet aggregation, which collectively contribute to improved heart health. Regular consumption of omega-3-rich foods, particularly fatty fish, has been associated with lower rates of heart attack, stroke, and other cardiovascular events.
2. **Anti-Inflammatory Properties**: The anti-inflammatory effects of omega-3 fatty acids are well-documented, making them valuable in managing chronic inflammatory conditions such as arthritis, inflammatory bowel disease, and metabolic syndrome. They help modulate inflammatory pathways, reducing the production of pro-inflammatory cytokines and mediators.
3. **Cognitive Function and Mental Health**: DHA, a key omega-3 fatty acid, is highly concentrated in the brain and plays a critical role in neuronal function and development. Research suggests that adequate intake of omega-3 fatty acids is associated with improved cognitive performance, better mood regulation, and a reduced risk of neurodegenerative diseases like Alzheimer's. Additionally, omega-3 supplementation has been linked to improvements in mental health conditions such as depression and anxiety.

4. **Vision Health**: DHA is also a major component of the retina, and adequate levels are essential for optimal visual function. Studies indicate that omega-3 fatty acids can help prevent age-related macular degeneration and other vision-related disorders.

The importance of omega-3 fatty acids in both fish nutrition and human health is profound and multifaceted. These essential nutrients are critical for supporting growth, reproduction, and immune function in fish, while also providing significant health benefits to humans. Given the increasing global demand for fish as a source of omega-3 fatty acids and the rising interest in their health benefits, understanding the role of these fatty acids in aquaculture and human nutrition is crucial for promoting both sustainability and health.

2. Understanding Omega-3 Fatty Acids

Omega-3 fatty acids are a specific group of polyunsaturated fatty acids (PUFAs) that play critical roles in various biological functions. Their unique chemical structure, characterized by the presence of multiple double bonds, imparts specific properties that are essential for the health of both aquatic and terrestrial organisms. Omega-3 fatty acids are classified into three main types:

2.1 Types of Omega-3 Fatty Acids

2.1.1 Alpha-Linolenic Acid (ALA)

Alpha-linolenic acid (ALA) is the simplest form of omega-3 fatty acid and is primarily found in plant sources. Key characteristics include:

- **Sources**: ALA is abundant in flaxseeds, chia seeds, walnuts, hemp seeds, and certain vegetable oils such as flaxseed oil and canola oil.
- **Essential Nature**: Unlike EPA and DHA, ALA is considered an essential fatty acid because the human body cannot synthesize it. Thus, it must be obtained through dietary intake.
- **Conversion**: ALA can be converted to EPA and DHA in the body, but the conversion rate is relatively low. This process is influenced by various factors, including age, gender, and overall health status.

2.1.2 Eicosapentaenoic Acid (EPA)

Eicosapentaenoic acid (EPA) is primarily derived from marine sources. It has several important characteristics:

- **Sources**: EPA is found in fatty fish (such as salmon, mackerel, and sardines), fish oils, and algae.
- **Biological Functions**: EPA is recognized for its role in anti-inflammatory processes and cardiovascular health. It serves as a precursor for the

synthesis of eicosanoids, which are signaling molecules that regulate inflammation, immune responses, and blood clotting.

2.1.3 Docosahexaenoic Acid (DHA)

Docosahexaenoic acid (DHA) is another crucial omega-3 fatty acid primarily found in marine sources. Its characteristics include:

- **Sources**: DHA is predominantly found in fish, fish oils, and marine algae.
- **Significance in the Brain**: DHA is a key structural component of neuronal membranes, particularly in the brain and retina. It plays an essential role in brain development, cognitive function, and vision health.
- **Health Benefits**: Like EPA, DHA is involved in anti-inflammatory processes and has been linked to improved cardiovascular health and brain function.

2.2 Essential Nature of Omega-3 Fatty Acids

The human body is incapable of synthesizing omega-3 fatty acids, particularly EPA and DHA. This makes them essential nutrients that must be sourced from dietary intake. Their crucial roles in cellular function, inflammation regulation, and overall health underscore the importance of including sufficient amounts of omega-3 fatty acids in both human and fish diets.

3. The Role of Omega-3 Fatty Acids in Fish Nutrition

Omega-3 fatty acids play vital roles in the growth, health, and overall performance of fish. Their significance is particularly pronounced in aquaculture, where optimizing nutrition can lead to improved production outcomes.

3.1 Importance in Fish Growth and Development

3.1.1 Cell Membrane Integrity

- **Structural Role**: Omega-3 fatty acids are integral components of cellular membranes. They contribute to the fluidity and permeability of membranes, which is essential for various cellular processes, including nutrient transport and signaling.
- **Physiological Impact**: The presence of omega-3 fatty acids in cell membranes supports optimal cellular function, enhancing the overall health and vitality of fish.

3.1.2 Growth Performance

- **Enhanced Growth Rates**: Adequate dietary levels of omega-3 fatty acids have been shown to improve growth rates in various fish species. For instance, fish receiving omega-3-rich diets tend to achieve larger sizes and better weight gain compared to those on low-omega-3 diets.
- **Feed Efficiency**: Omega-3 fatty acids also enhance feed conversion efficiency, allowing fish to utilize feed more effectively for growth, ultimately improving aquaculture productivity.
- **Muscle Development**: Supplementing fish diets with omega-3 fatty acids has been linked to better muscle development and quality, which is important for market value and consumer preferences.

3.1.3 Reproductive Health

- **Gamete Quality**: Omega-3 fatty acids are crucial for the quality of gametes in broodstock fish. Diets enriched with EPA and DHA can lead to improved egg and sperm quality, which in turn affects fertilization success.
- **Embryo Development**: The presence of omega-3 fatty acids in the diet influences embryo development, contributing to higher survival rates of larvae during critical early life stages.
- **Larval Survival**: Incorporating omega-3 fatty acids in larval diets supports immune function and overall health, enhancing survival rates during the vulnerable larval stage.

3.2 Immunological Benefits

Fish are frequently exposed to stressors such as pathogens, fluctuating water quality, and environmental changes. Omega-3 fatty acids help bolster the immune system in several ways:

3.2.1 Enhancing Immune Function

- **Immune Modulation**: EPA and DHA can modulate the activity of immune cells, enhancing their ability to respond to infections and pathogens. Studies have shown that fish with higher dietary omega-3 levels exhibit improved immune responses to vaccinations and reduced susceptibility to diseases.
- **Increased Resistance**: By enhancing immune cell function, omega-3 fatty acids help fish resist infections and manage stressors more effectively.

3.2.2 Reducing Inflammation

- **Anti-Inflammatory Properties**: Omega-3 fatty acids possess potent anti-inflammatory properties. They help mitigate the effects of chronic inflammation, which can compromise fish health and performance.
- **Improved Health Outcomes**: By reducing inflammation, omega-3 fatty acids contribute to overall fish health, enabling better recovery from stress and disease.

3.3 Cardiovascular Health in Fish

The cardiovascular system of fish is crucial for maintaining health and performance, particularly under stressful conditions. Omega-3 fatty acids support cardiovascular health in several ways:

3.3.1 Blood Circulation

- **Improved Circulation**: Omega-3 fatty acids enhance blood circulation by reducing platelet aggregation and promoting vasodilation. This is essential for maintaining optimal blood flow and nutrient delivery to tissues.
- **Nutrient Transport**: Enhanced circulation supports the effective transport of nutrients and oxygen to cells, contributing to overall metabolic health.

3.3.2 Heart Function

- **Enhanced Cardiac Function**: Dietary omega-3 fatty acids can improve heart function, reducing the risk of cardiac stress during periods of environmental stress, disease outbreaks, or other challenges.
- **Stress Resilience**: By supporting heart health, omega-3 fatty acids help fish cope with stressors more effectively, improving overall resilience and survival rates in aquaculture settings.

Understanding the various roles of omega-3 fatty acids in fish nutrition is essential for optimizing aquaculture practices and promoting fish health. By ensuring adequate dietary levels of these critical nutrients, aquaculture producers can enhance growth performance, reproductive success, immune function, and cardiovascular health in fish. The implications of these benefits extend beyond fish health, as omega-3 fatty acids also hold significant promise for human health, highlighting the interconnectedness of aquatic and human nutrition. In the subsequent sections, we will delve deeper into the metabolic pathways of omega-3 fatty acids, their sources, and practical applications in aquaculture nutrition.

3. The Role of Omega-3 Fatty Acids in Fish Nutrition

Omega-3 fatty acids are essential nutrients that play a critical role in the nutrition and overall health of fish. They are especially important in aquaculture, where optimizing fish growth, health, and productivity is paramount. This section delves into the various ways omega-3 fatty acids influence fish nutrition, including their roles in growth and development, immunological benefits, and cardiovascular health.

3.1 Importance in Fish Growth and Development

3.1.1 Cell Membrane Integrity

- **Structural Components**: Omega-3 fatty acids are integral to the phospholipid bilayer of cell membranes. Their unique chemical structure, characterized by multiple double bonds, contributes to the fluidity and flexibility of the membrane. This fluidity is crucial for maintaining the integrity of the cell, allowing for proper nutrient transport and cell signaling.
- **Nutrient Transport**: The structural integrity provided by omega-3 fatty acids facilitates the effective transport of essential nutrients into the cell and waste products out of the cell. This balance is essential for maintaining cellular homeostasis and function.
- **Cell Signaling**: Omega-3 fatty acids influence cell signaling pathways that regulate various physiological processes, including growth, metabolism, and immune responses. Their presence in the cell membrane can affect receptor activity, thus impacting overall cellular function.

3.1.2 Growth Performance

- **Enhanced Growth Rates**: Research has consistently shown that dietary omega-3 fatty acids, particularly EPA and DHA, significantly enhance growth rates in various fish species. Fish that are provided with adequate omega-3 levels often exhibit better weight gain compared to those fed diets deficient in these fatty acids.
- **Feed Efficiency**: Omega-3 fatty acids contribute to improved feed conversion ratios (FCR) by enhancing the efficiency with which fish utilize feed for growth. This is particularly important in aquaculture, where feed costs are a significant factor in production economics.
- **Muscle Development**: Omega-3 fatty acids promote muscle growth and development in fish. Adequate levels in the diet lead to increased muscle fiber size and number, which contributes to the overall quality of the fish and enhances market value.

- **Gamete Quality**: Omega-3 fatty acids are critical for the reproductive performance of fish. They influence the quality of eggs and sperm, with diets rich in EPA and DHA improving the viability and fertilization rates of gametes.
- **Embryo Development**: The presence of omega-3 fatty acids in the diet supports healthy embryo development. This is crucial for ensuring high survival rates during the early life stages of fish, a period often characterized by high mortality rates.
- **Larval Survival**: Omega-3 fatty acids are vital for larval nutrition, influencing not only their growth but also their immune competence and resilience to environmental stressors. Adequate omega-3 levels in larval diets lead to better survival rates during the critical first weeks of life.

3.2 Immunological Benefits

Fish in aquaculture settings are frequently exposed to pathogens, environmental stressors, and challenges that can compromise their immune systems. Omega-3 fatty acids offer several immunological benefits:

3.2.1 Enhancing Immune Function

- **Modulation of Immune Cells**: EPA and DHA play an important role in modulating the activity of immune cells, enhancing their ability to respond to infections and diseases. This is crucial for maintaining fish health, especially in densely populated aquaculture environments where the risk of disease transmission is high.
- **Resistance to Pathogens**: Fish with diets enriched with omega-3 fatty acids tend to exhibit increased resistance to various pathogens, leading to lower incidence rates of diseases. This can significantly reduce the need for antibiotics and other treatments in aquaculture.

3.2.2 Reducing Inflammation

- **Anti-Inflammatory Properties**: Omega-3 fatty acids are well-known for their anti-inflammatory properties. They help reduce the production of pro-inflammatory cytokines and eicosanoids, which can mitigate chronic inflammation within the fish.
- **Support for Overall Health**: By reducing inflammation, omega-3 fatty acids support overall fish health, allowing them to recover more quickly from stressors and minimizing the impact of diseases.

3.3 Cardiovascular Health in Fish

The cardiovascular system is vital for maintaining overall fish health, particularly during stressful conditions. Omega-3 fatty acids contribute to cardiovascular health in various ways:

3.3.1 Blood Circulation

- **Improved Circulation**: Omega-3 fatty acids enhance blood circulation by reducing platelet aggregation and promoting vasodilation. These effects help maintain optimal blood flow, which is essential for delivering nutrients and oxygen to tissues.
- **Nutrient Delivery**: Effective circulation ensures that nutrients absorbed from the diet are efficiently transported to all parts of the fish's body, supporting metabolic functions and growth.

3.3.2 Heart Function

- **Enhanced Cardiac Function**: Dietary omega-3 fatty acids can improve heart function, which is particularly important during periods of environmental stress or disease outbreaks. The presence of omega-3s can help stabilize heart rhythms and reduce cardiac stress.
- **Stress Resilience**: By supporting heart health, omega-3 fatty acids contribute to the resilience of fish under stressful conditions. This enhances survival rates and overall productivity in aquaculture settings.

4. Human Health Benefits of Omega-3 Fatty Acids

Beyond their importance in fish nutrition, omega-3 fatty acids are recognized for their numerous health benefits in humans. This section outlines the key benefits associated with omega-3 fatty acids.

4.1 Cardiovascular Health

Numerous studies have demonstrated the significant cardiovascular benefits of omega-3 fatty acids:

- **Cholesterol Regulation**: Omega-3 fatty acids can lower triglyceride levels while increasing high-density lipoprotein (HDL) cholesterol, which is associated with a reduced risk of heart disease. This lipid profile improvement is crucial for cardiovascular health.
- **Blood Pressure Reduction**: Regular consumption of omega-3 fatty acids has been linked to lower blood pressure levels. This effect contributes to overall cardiovascular health and reduces the risk of hypertension-related complications.

- **Anti-Arrhythmic Effects**: Omega-3 fatty acids may help prevent arrhythmias and other heart rhythm disorders, enhancing cardiovascular safety. This property is especially important for individuals with existing heart conditions.

4.2 Anti-Inflammatory Properties

Omega-3 fatty acids exhibit potent anti-inflammatory effects, making them beneficial for various inflammatory conditions:

- **Chronic Inflammation**: Omega-3 fatty acids can alleviate symptoms associated with chronic inflammatory diseases, such as arthritis and inflammatory bowel disease. Their ability to modulate inflammatory pathways helps in managing these conditions effectively.
- **Autoimmune Disorders**: Omega-3 supplementation may improve symptoms of autoimmune disorders by modulating the immune response and reducing inflammation, leading to better disease management and quality of life for affected individuals.

4.3 Brain Health

The impact of omega-3 fatty acids on brain health is profound and multifaceted:

- **Cognitive Function**: DHA is a primary structural component of the brain and is essential for maintaining cognitive function. Adequate intake of DHA is associated with improved memory and learning capabilities, particularly in aging populations.
- **Mental Health**: Studies suggest that omega-3 fatty acids may help alleviate symptoms of depression and anxiety, promoting better mental health. Their role in brain health underscores the importance of these fatty acids for emotional and cognitive well-being.

5. Dietary Sources of Omega-3 Fatty Acids

Omega-3 fatty acids, particularly EPA and DHA, are vital for both fish and human health. They can be sourced from various dietary options, with significant differences in the types and concentrations of omega-3s available. This section explores the primary dietary sources of omega-3 fatty acids, categorized into marine sources, plant sources, and algal oil.

5.1 Marine Sources

Marine sources are the most concentrated and bioavailable sources of EPA and DHA, making them crucial for both aquaculture and human diets.

5.1.1 Fatty Fish

Fatty fish are renowned for their high omega-3 content. Consuming these fish regularly can significantly boost dietary omega-3 levels:

- **Salmon**: Particularly rich in EPA and DHA, salmon is a highly recommended source. Both wild-caught and farmed salmon offer substantial omega-3 levels, although wild varieties may have slightly higher concentrations.
- **Mackerel**: This oily fish is another excellent source of omega-3 fatty acids. Atlantic mackerel is especially high in EPA and DHA and is often featured in Mediterranean diets.
- **Sardines**: Small but mighty, sardines are packed with omega-3s and are often consumed canned. They are also a sustainable choice, making them a popular option for health-conscious consumers.
- **Herring**: Similar to sardines, herring is an oily fish that provides a rich supply of omega-3 fatty acids. It can be enjoyed smoked, pickled, or fresh.
- **Trout**: Rainbow trout, particularly those raised in clean, cold waters, can be a great source of omega-3s. They are often farmed in sustainable practices, ensuring their availability.

5.1.2 Fish Oil Supplements

- **Concentrated Source**: Fish oil capsules offer a concentrated form of EPA and DHA, often derived from fatty fish. These supplements can help individuals meet their omega-3 dietary requirements, particularly for those who do not consume enough fish.
- **Health Benefits**: Regular intake of fish oil supplements has been associated with numerous health benefits, including improved cardiovascular health and enhanced cognitive function. It is essential to choose high-quality supplements that are free from contaminants, such as heavy metals and toxins.

5.2 Plant Sources

While plant sources predominantly contain ALA, they are still valuable contributors to overall omega-3 intake. ALA can be converted into EPA and DHA in the body, although this conversion is often inefficient.

5.2.1 Flaxseeds

- **High in ALA**: Flaxseeds are one of the richest plant sources of omega-3 fatty acids, particularly ALA. They can be consumed whole, ground, or as flaxseed oil.

- **Nutritional Benefits**: Flaxseeds are also rich in fiber, lignans, and antioxidants, providing additional health benefits. They can be easily added to smoothies, baked goods, or salads for a nutritional boost.

5.2.2 Chia Seeds

- **Another ALA Source**: Chia seeds are another excellent plant-based source of omega-3 fatty acids. They can absorb liquid and form a gel-like consistency, making them a versatile ingredient in various dishes.
- **Incorporation into Diet**: Chia seeds can be added to smoothies, oatmeal, yogurt, and baked goods, enhancing their omega-3 content while also providing fiber and protein.

5.2.3 Walnuts

- **Rich in ALA**: Walnuts provide a healthy dose of ALA and are a convenient snack option. They are versatile and can be added to salads, snacks, and baked goods.
- **Nutritional Profile**: In addition to omega-3s, walnuts are rich in antioxidants, vitamins, and minerals, contributing to overall health.

5.3 Algal Oil

For individuals following vegetarian or vegan diets, algal oil supplements derived from microalgae offer a direct source of DHA and EPA, making them an excellent alternative to fish-derived sources.

- **Direct Source of EPA and DHA**: Algal oil provides a plant-based source of the same beneficial omega-3 fatty acids found in fish, making it a suitable option for those who do not consume animal products.
- **Sustainability**: Algal oil production is considered more sustainable than fish oil harvesting, as it reduces pressure on fish populations and marine ecosystems. This makes algal oil a compelling choice for environmentally conscious consumers.

6. Nutritional Considerations in Aquaculture

In aquaculture, optimizing omega-3 fatty acid levels in fish diets is essential for promoting both fish health and human health. This section discusses the importance of diet formulation, sustainable sourcing, and consumer education regarding omega-3 fatty acids.

6.1 Diet Formulation

- **Optimizing Nutritional Profiles**: Formulating fish diets with adequate levels of omega-3 fatty acids ensures that farmed fish provide a beneficial source of these nutrients to consumers. Diets can be designed to incorporate various sources of omega-3s, including fishmeal, fish oil, and plant-based alternatives.

- **Balancing Nutritional Needs**: Achieving a balance between omega-3 and omega-6 fatty acids is crucial for fish health. Proper formulation helps maximize growth, health, and feed efficiency in aquaculture systems.

6.2 Sustainable Sourcing

- **Sourcing Omega-3s**: With the increasing demand for omega-3 fatty acids, sustainable sourcing of fishmeal and fish oil is crucial to meet both aquaculture and human dietary needs. Practices such as responsible fishing and aquaculture techniques can help ensure the long-term availability of omega-3 sources.
- **Plant-Based Alternatives**: Developing plant-based alternatives to fishmeal and fish oil is essential for reducing reliance on marine resources while maintaining high omega-3 levels in aquaculture feeds. Innovations in feed technology can lead to more sustainable aquaculture practices.

6.3 Health Claims

- **Consumer Education**: Educating consumers about the health benefits of omega-3-rich fish products can promote better dietary choices and increase the consumption of omega-3 fatty acids. Clear labeling and marketing strategies can help highlight the nutritional advantages of omega-3s in fish.
- **Promoting Dietary Choices**: By emphasizing the health benefits of omega-3s, aquaculture can play a role in encouraging consumers to incorporate more fish and omega-3-rich products into their diets, ultimately supporting public health initiatives.

7. Conclusion

Omega-3 fatty acids play a crucial role in fish nutrition, promoting growth, immune function, and overall health. Their importance extends to human health, where they contribute to cardiovascular health, anti-inflammatory responses, and cognitive function. With the rising awareness of the health benefits associated with omega-3 fatty acids, both aquaculture and human dietary strategies must prioritize the incorporation of these essential nutrients. Sustainable practices in fish farming and nutritional formulations will ensure a continued supply of omega-3 fatty acids, benefiting both aquatic species and human consumers alike. As the demand for healthy, omega-3-rich diets increases, the aquaculture industry must adapt to provide high-quality, sustainable sources of these vital nutrients, thereby supporting both human and environmental health.

References

Bó, D. M., et al. (2017). Omega-3 fatty acids in aquaculture: A review of current knowledge. Aquaculture, 473, 161-173.

Brenna, J. T., et al. (2009). Omega-3 fatty acid supplementation: A review of the evidence. Nutrition Reviews, 67(5), 271-284.

Brett, J. R. (1979). Environmental factors and growth of salmonids. Journal of the Fisheries Research Board of Canada, 36(3), 365-371.

Calder, P. C. (2015). Omega-3 fatty acids and inflammatory processes: From molecules to man. Biochemical Society Transactions, 43(5), 1010-1014.

DeWitt, W. R., et al. (2017). The role of dietary omega-3 fatty acids in human health. American Journal of Lifestyle Medicine, 11(3), 234-242.

Dyerberg, J., & Bang, H. O. (1979). Fatty acid composition of the plasma lipids in Greenland Eskimos. American Journal of Clinical Nutrition, 32(6), 1911-1919.

Gonzalez-Gallego, J., et al. (2012). Omega-3 fatty acids: The role of diet and health. World Journal of Gastroenterology, 18(32), 4337-4346.

Gonzalez-Navarro, B., et al. (2017). Fatty acids in aquaculture: The case for omega-3 and omega-6 fatty acids. Aquaculture Reports, 5, 135-146.

Kris-Etherton, P. M., & Hecker, K. D. (2002). Fish consumption, fish oil, omega-3 fatty acids, and cardiovascular disease. Nutrients, 14(1), 31-63.

Kris-Etherton, P. M., et al. (2004). Fish consumption, fish oil, omega-3 fatty acids, and cardiovascular disease. American Journal of Clinical Nutrition, 80(6), 1612-1619.

Liu, M., et al. (2015). Fish oil supplementation and cardiovascular risk: A systematic review and meta-analysis. American Journal of Clinical Nutrition, 102(6), 1568-1579.

McManus, M. (2008). The role of omega-3 fatty acids in brain health. Nutrition Reviews, 66(6), 380-396.

Olsen, R. E., & Holmer, M. (2015). Aquaculture and human nutrition: A review of the role of omega-3 fatty acids. Nutrients, 7(7), 6758-6774.

Pérez, J. A., et al. (2021). The role of omega-3 fatty acids in human health: A review. Health, 13(5), 725-731.

Simopoulos, A. P. (1999). Essential fatty acids in health and chronic disease. American Journal of Clinical Nutrition, 70(3), 560-569.

Turchini, G. M., & Trushenski, J. (2009). The role of omega-3 fatty acids in fish nutrition. Reviews in Fish Biology and Fisheries, 19(2), 155-170.

Zhou, Q., et al. (2019). The role of omega-3 fatty acids in fish nutrition and health: A review. Aquaculture Nutrition, 25(3), 579-591.

Index